PRE-FEASIBILITY STUDY OF CROP AND SMALL AND MEDIUM-SIZED ENTERPRISE INSURANCE PILOTS IN NEPAL'S MUNICIPALITIES

APRIL 2025

ASIAN DEVELOPMENT BANK

CONTENTS

TABLES, FIGURES, AND BOXES

ACKNOWLEDGMENTS

This report was prepared under the technical assistance *Strengthening the Enabling Environment for Disaster Risk Financing (Phase 2)*, a project executed by the Asian Development Bank (ADB).

Arup Chatterjee, principal financial sector specialist, Finance Sector Office, Sectors Group, ADB provided direction and technical advice.

The report was produced by a team of ADB consultants comprising Arindom Baidya, agriculture insurance specialist (international consultant); Mayank Dubey, insurance and disaster risk specialist (international consultant); Udaya Raj Adhikari, insurance industry specialist (national consultant); and Maria Cristina Pascual, project coordinator (national consultant).

The report benefited extensively from the generous participation of many government agencies, private sector organizations, and development partners in meetings and workshops organized under the project. Special thanks to Malay Poddar (former chairman and managing director, Agriculture Insurance Company of India Ltd.); Anuj Kumbhat (chief executive officer, Weather Risk Management Services, India); and Swapnil Sone (agriculture underwriter, SCOR Re, Zurich) for participating in the workshop organized in Kathmandu. The report team expresses great appreciation to the staff of these entities for their time and candid opinions.

ABBREVIATIONS

ADB	Asian Development Bank
AWS	automated weather station
CLID	Crop and Livestock Insurance Directive
FY	fiscal year
GDP	gross domestic product
MSMEs	micro, small and medium-sized enterprises
NCIP	Nepal Catastrophe Insurance Pool
SMEs	small and medium-sized enterprises

EXECUTIVE SUMMARY

Crop insurance and small and medium-sized enterprise (SME) insurance are important in government efforts to reduce contingent financing needs in disasters. It is equally important to the financial resilience of individual farmers and business owners to get financial protection against disaster-induced loss.

The government can source finance to recover financial and economic losses through varied sources. But individuals do not have that advantage: they can only "transfer" (manage) such risk with insurance. Unfortunately, traditional insurance has been unable to reach individual farmers and SMEs in many places, and alternative solutions are needed within risk transfer frameworks. This study identifies shortcomings and proposes a high-level framework that *can* protect farmers and SMEs. The study assesses alternative insurance mechanisms for crop insurance and SMEs at the sub-sovereign level. Existing crop and livestock insurance in Nepal requires better products and processes in the short term and a more sustainable platform for the longer term.

Yet, the SMEs and farmers could significantly mitigate the impact of natural hazards with effective disaster risk financing solutions. Insurance is nonetheless not a stand-alone risk financing solution. Its suitability needs to be assessed based on criteria such as the frequency and severity of the hazards faced, the opportunity cost of premium payments, and the availability of alternative mechanisms for managing risk. The selection process should involve assessing the scale of risk at each return period, determining the ownership of risk, and identifying the most suitable instrument for each layer of risk. A thorough risk assessment and cost-benefit review are critical to matching risk financing options. The selection of sovereign risk transfer instruments is ultimately a political exercise that requires balancing timeliness, accuracy, and cost, and establishing an enabling environment for private sector participation.

Technical assistance for the project *Strengthening the Enabling Environment for Disaster Risk Financing (Phase 2)* is focused on feasibility studies on disaster insurance products for mitigating agriculture risk, restoring livelihoods, and helping homeowners and small businesses in Nepal. The technical assistance recommends potential insurance pilots for post-disaster restoration of livelihoods of small and marginal farmers and small businesses in municipalities in Nepal. For the feasibility study, the municipalities of Birgunj, Bhimeshwor, and Janakpur were considered. However, the pilots will make the final selection of municipalities in consultation with the Government of Nepal.

A couple of study missions and workshops were organized, focused on understanding the feasibility of weather index-based crop insurance against losses of yield for small and marginal rice farmers who had borrowed from financial institutions. They also aimed to assess post-disaster assistance to homeowners and small businesses using parametric index insurance products linked to social assistance schemes. In addition, they explored cost-sharing arrangements for insurance premiums among federal, provincial, and local governments, and the involvement of private insurance companies.

The missions and workshops aimed to do the following:

(i)   Meet key stakeholders. These include municipal governments; farmers; SMEs and micro, small, and, medium-sized enterprises associations; federal government departments; the insurance regulator; key agencies collecting and analyzing weather and hydro metrological risk data; the National Disaster Risk Reduction and Management Authority; insurers; and reinsurers to understand readiness to use parametric insurance instruments for risk transfer at the sub-sovereign level.

(ii) Learn from parametric pilot initiatives promoted by development partners.

(iii) Provide potential options for financing disaster risk efficiently at the sub-sovereign level to enhance financial preparedness.

(iv) Arrive at cost estimates for potential pilots for parametric crop insurance, and SMEs' and homeowners' insurance.

Main challenges:

- The government has adopted measures to support agriculture by providing concessional credit to the agriculture sector and subsidizing agriculture insurance by up to 80%. However, the benefits of these measures have not yet percolated down, as banks are reluctant to lend to farmers due to the inherent risks and insurance companies promoting livestock insurance over crop insurance.

- Reluctance from insurance companies to underwrite crop insurance. This is not just because of inadequate premiums but also because of other structural problems, such as premium subsidies not being received on time and claims assessment-related anomalies.

- Operational difficulties include terrain, lack of human resources (especially surveyors), procedural difficulties in enrollment, and claims handling.

- Lack of awareness, proper promotion, financial literacy, and trust in insurance.

- Dissatisfaction with the estimation of input cost as the basis of indemnity for crop insurance. Lack of transparency in sum insured calculations.

- Missing technology usage.

- Indemnity-based products have serious disadvantages for small farmers:
  - cumbersome procedures,
  - high administrative costs,
  - lengthy claims settlement,
  - transparency,
  - moral hazard, and
  - challenging to scale due to high operational costs.

- Existing parametric insurance schemes provide a claims payout only when triggered. If the trigger is not reached, policy holders do not receive any payout, even if they have suffered a slight loss.

- Lack of structured data for modeling parametric risk and pricing, as well as inadequate weather-related infrastructure and data analytics systems and capabilities for claims settlement.

Proposed solutions:

- Carry out parametric pilots for rice crops in two locations against weather risks—drought, rainfall, wind, and flood.

- Carry out pilots for small businesses' and homeowners' insurance to provide rapid payout post disaster, for flood and earthquake risks, to affected residents as cash transfers for restoration of livelihoods.

- Begin pilots with 500 farmers and SMEs, with numbers to grow gradually.

- Parametric solutions will involve less administrative cost, no moral hazard, and assuring timely disbursement of claims.

- Involvement of local government bodies is important for scalability and sustainability. Municipalities can act as risk aggregators and will make distribution and disbursement easier.

- Both these solutions will help provide municipalities with fiscal space to manage expenses post disaster.

- Set up a risk pool to make the product affordable and scalable. This will be an efficient mechanism wherein the municipalities pool their risk and can reduce premiums due to the diversification benefits—across geography, time, and portfolio.

- The risk pool will retain a specific portion of the risk and use it to make payments for smaller, more frequent claims (e.g., 25% to 75%) when no trigger is hit. A portion of the funds may also be used for capacity building and employing staff to carry out specific activities. The arrangement will ensure that municipalities represented in the pool do not cross-subsidize.

- The risk pool will provide scale and negotiation strength.

- The pool will be managed by a team of experts, including representation from the Ministry of Finance, Ministry of Agriculture, Nepal Insurance Authority, municipalities, the National Disaster Risk Reduction and Management Authority, insurance companies, reinsurance companies, and other industry experts.

- Linking concessional farm credit offered by governments with insurance through banks and microfinance institutions is being considered so the lender can insure the entire loan book. Such an arrangement may give banks greater confidence to lend and restructure loans post disaster.

- Government should review and evaluate agriculture credit and its accessibility to farmers. Small farmers are deprived of credit during the sowing season, and subsidized credit to farmers needs a capping of the upper limit of up to NRs1 million, depending on the crop. Accessing credit during the sowing season and repayment during the harvest is important for small farmers; this can lead to better use of financial assistance and protection during disasters.

- The product will build incentives to encourage joint bank accounts, with one female family member to pay a share of the premium and receive claims.

- Introduce women as *Beema Bahinis*[1] for insurance literacy, documentation, onboarding of farmers, and monitoring by equipping them with smartphones.

- Develop smartphone apps for enrolling and managing farmers and providing weather advisories.

- The government subsidy is essential for the insurance premium. The subsidy can be reduced over time. International experience shows that agriculture insurance subsidies are crucial to success; otherwise, scaling up will be difficult.

There is general support for this approach from the Nepal Insurance Authority, insurers and reinsurers, and federal government agencies. Municipal governments have also expressed support. Nepal Re and Himalayan Re have assured reinsurance support and managed the pool pro bono during the initial setup. Proposed pilots will enhance awareness, financial literacy, and trust in insurance and fulfill the primary objectives of disaster risk financing.

---

[1] "Beema Bahini" is a term coined for the project, where Beema translates to "insurance" and Bahini translates to "sister" in Nepali, symbolizing a community-oriented approach to insurance initiatives where young women lead it.

Way Forward

It is evident that recommended pilots for small and marginal farmers, SMEs, and homeowners will be an effective risk transfer tool to accomplish disaster risk management objectives. This observation is based on this study and discussions with government departments, the insurance regulator, municipalities, farmers, insurance companies, multilateral organizations, and other industry experts. The following measures may contribute to better implementation and scalability of the proposed pilot.

- Insurance-related regulations are evolving in Nepal. Regulatory issues, including approval of products and setting up the insurance risk pool, are well practiced within the supervision of the insurance regulator, and risk-specific pools can be formed and should be regulated.

- Stakeholders' goals are not aligned,[2] and each stakeholder has different expectations and apprehensions.[3] It is essential that the expectations of all major stakeholders—all three levels of government, insurance sectors, and financial institutions—are documented and addressed. This can be achieved by active campaigning, conducting workshops, and using other similar tools. Active participation of multilateral organizations, research organizations, industry experts, and nongovernment organizations could help achieve objectives. Agriculture insurance literacy programs and training will increase awareness among farmers.

- Capacity-related gaps are also a significant issue to work on. Stakeholders in Nepal are not fully equipped with the technical and operational aspects of disaster and crop insurance. Concepts such as parametric insurance and the use of remote sensing[4] for insurance are still new. It will be good to (i) build a deeper appreciation for the scope, technical possibilities, and challenges of building disaster insurance products among policymakers, and (ii) create capacity for robust product design and smart underwriting within the insurance and associated sectors.

- Lack of quality data is a serious issue. Weather and satellite data will be required to help price the risks and provide robust triggers and premium rates for parametric solutions. Government agencies such as the National Disaster Risk Reduction and Management Authority, the Ministry of Geology, the Ministry of Agriculture, the Meteorological Department, and multilateral agencies may help provide necessary data. Automatic weather stations can also be established to fulfil the additional need. Collaboration between agencies is important in achieving real-time quality data.

- Premium subsidies are essential for the success and scalability of the pilot. Potential sharing of premiums among federal, provincial, municipal, individual businesses, and households, and funding from multilateral organizations, needs to be worked out. Timely transfer of the subsidy is essential.

- Take-up rates of agriculture insurance in Nepal are very low. Active participation of municipal governments is important for scalability and sustainability. Municipalities can act as risk aggregators and will make distribution and disbursement easier. The proposed pilot will provide a platform for municipalities to utilize disaster relief funds more effectively and will fulfil the social commitment of the Government of Nepal. Communication among the stakeholders will be better and more effective due to the participation of the municipal government. Other operational challenges, such as high administrative costs, may also be more easily addressed with their involvement.

---

2   For example, the Ministry of Agriculture has decided not to provide blanket subsidy support to crop insurance pilots. This is probably due to a paucity of funds.

3   Insurance companies in Nepal are quite apprehensive about crop insurance because (i) premium is low, (ii) operational costs are high, and (iii) government does not release premium subsidies.

4   A few organizations, such as the International Centre for Integrated Mountain Development and technical governmental agencies, have excellent technical knowledge but do not have experience in parametric insurance.

Municipalities can build up reserves for catastrophic events when disasters do not happen or use the collected premium for strategic spending for maintenance and risk literacy.

- Several farmers have expressed dissatisfaction with the estimation of input cost as the basis of indemnity for crop insurance, which is cited as a major cause of their reluctance. A transparent, well-defined, and acceptable procedure should be developed to determine the sum insured.

- A tariff-based approach is not ideal, and premium rates should be redetermined on an actuarial basis for each district-season-crop combination, given the difference in risk exposures. This will help increase the participation of insurance companies and provide reasonable comfort to reinsurance companies.

- Since the proposed solutions are based on a parametric insurance approach, it is necessary to eliminate basis risk. Proper calibration of triggers and exits is essential. High basis risk may create serious dissatisfaction among stakeholders and may eventually create serious reluctance toward the product.

- Careful creation and effective management of the proposed disaster insurance pool are equally important. The pool needs initial capitalization. The Government of Nepal should explore options for sourcing the finance internally or externally. International donors like the World Bank, Swiss Agency for Development Corporation, International Finance Corporation, and KfW might also be willing to seed the pool. The pool also needs strong political commitment to succeed. Strong intent and coordination between the Government of Nepal and municipalities have to be demonstrated during the initial phases to attract full stakeholder interest. Part of disaster risk finance funds allocated at various levels can also be utilized for initial funding of the pool.

- It would be optimal if various organizations came together on a common platform for the pilot and subsequently became part of the proposed disaster pool. Explore potential partnerships with organizations such as the International Centre for Integrated Mountain Development and development partners such as the United Nations Capital Development Fund, Germany's GIZ, etc. This will increase pilot efficiency and reduce costs.

- Use of the latest technology will further reduce costs and help enhance take-up rates. A digital platform can be created to manage the entire ecosystem. Smartphone apps and web-based technologies can be used to automate insurance-related formalities. Smartphones can also be used for communication, e.g., alerts such as weather-related advisories.

- As a policy initiative, farm loans can be linked with insurance policies. Also, incentives can be extended to encourage joint bank accounts, with one female family member paying a share of the premium and receiving claims.

- To address the gender issue, female volunteers (*Beema Bahinis*) can be equipped with smartphones and trained as technical assistants for insurance literacy, documentation, onboarding of farmers, and monitoring.

1

Context and Scope of the Study

This report presents the observations and conclusions of (i) two missions to Nepal, to the municipalities of the Birgunj, Janakpur, and Bhimeshwor, and (ii) discussions with various stakeholders nationally. The study analyzes crop insurance and sub-sovereign insurance as a tool to strengthen disaster risk financing mechanisms.

In an earlier study, government officials, policymakers, and insurance regulators expressed the need to strengthen Nepal's financial management of disaster risk to smooth the cost of disasters and ensure timely post-disaster funding (ADB 2019a).

Nepal's federal governance structure allows decentralized disaster risk management, with potential benefits of location-specific structural risk reduction, post-disaster events, and quick cash/in-kind support to communities for relief and rehabilitation. This follow-up Phase 2 study focuses more on developing an operational framework for an integrated risk financing strategy in a risk-layered approach.

The report (i) reviews disaster and crop insurance in Nepal, (ii) recommends suitable insurance products and structures, and (iii) attempts to define the contours of a feasibility study, followed by a pilot (on crop and disaster insurance).

Nepal's Geography, Climate, and Risk Profile

Nepal is a small and landlocked Himalayan country in South Asia, which nonetheless features three distinct geographical regions, including the Terai in the south, the Middle Hills in the center, and the Himalayas in the north. The agricultural plainland of the Terai covers about 17% of the total area and is an extension of the Ganges Plain. Agriculturally, it is the most productive area and important for food security. The Middle Hills region covers about 68% of the total area. The Himalayas are the world's highest mountains and cover about 15% of the total. The population here is thin, and snow covers most of the region's ground throughout the year.

Mean annual temperature varies with elevation and the gradient from north to south. In the Himalayas, mean annual temperatures are well below 0°C degrees, and in the Terai, it is often over 24°C.

The distribution of precipitation in Nepal is very complex due to topographic variation. In areas of heavy precipitation, annual precipitation ranges from 3,500 millimeters to 5,000 millimeters. Around the city of Pokhara, average precipitation is over 5,000 millimeters annually.

Nepal is extremely disaster-prone due to its topography and climatic conditions. The country is exposed to a wide range of natural hazards, including floods, glacial lake outburst floods, landslides, windstorms, hailstorms, droughts, and earthquakes. It is located in a seismically active zone with a high probability of major earthquakes. It ranks fourth in its relative vulnerability to climate change and 11th in earthquakes globally.

Climate change is already significantly impacting Nepal's environment, with species' ranges shifting to higher altitudes, glaciers melting, and the frequency of precipitation extremes increasing from June to September. Average maximum precipitation is 300 millimeters, and Figure 1 shows the relationship between precipitation and temperature. A joint report by the World Bank and ADB (2021) makes the following points:

(i)   Warming in Nepal is projected to be higher than the global average. By the 2080s, Nepal could warm by 1.2°C–4.2°C under the highest emission scenario.

(ii) Natural hazards, such as drought, heatwave, river flooding, and glacial lake outburst flooding, are all projected to intensify over the 21st century.

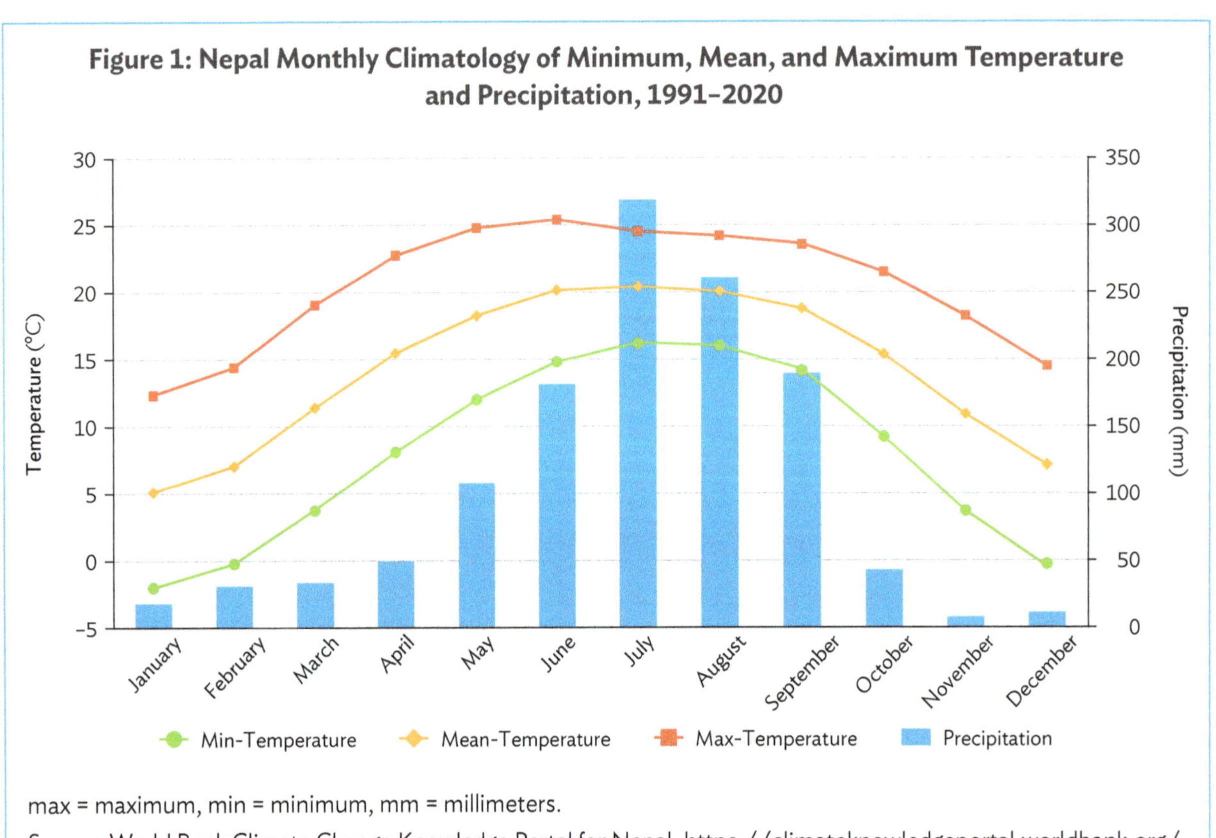

Figure 1: Nepal Monthly Climatology of Minimum, Mean, and Maximum Temperature and Precipitation, 1991–2020

max = maximum, min = minimum, mm = millimeters.

Source: World Bank Climate Change Knowledge Portal for Nepal. https://climateknowledgeportal.worldbank.org/country/nepal/climate-data-historical (accessed 19 July 2023).

(iii) Modeling has suggested that the number of people annually affected by river flooding could more than double by 2030 due to climate change. At the same time, the economic impact of river flooding could triple.

(iv) Based on Government of Nepal estimates, 1.9 million people are highly vulnerable to the potential impacts of climate change, while an additional 10 million are increasingly at risk. The economic cost of climate change to agriculture, hydropower, and industries affected by water-induced disasters is estimated to equal 1.5%–2.0% of gross domestic product (GDP) per year by 2050.

Socioeconomics

Among Nepal's 29.2 million people, about 48.87% are male, and 51.13% female, and the population is growing at a rate of 0.92%. The country's population density is 198 per square kilometer, with the highest population density of 460 in the Terai region (CBS 2021). It is a young country, with a population below 14 years of age at 28.83% and 15–59 years at 61.96%. As per the 2021 census, 53.61% of the population lives in the Terai, 40.31% in the Middle Hills, and 6.08% in the Mountain region (CBS 2021).

About 15 million people are engaged in economic activities, with (i) 50.1% engaged in agriculture, forestry, and fisheries; (ii) 23.0% as elementary (unskilled) workers; and (iii) the rest in professional services, including the armed forces (CBS 2021).

About 6.6 million households exist in the country, with an average of 4.37 people per household: 23.8% of households own land or a housing unit, or both, in the name of a female household member. The census results show that 9.4% of households operate small-scale enterprises other than agriculture and have no paid employees. Among such households, 137,644 (21.9%) operate a cottage industry; 310,651 (49.5%) operate in trade/business; 34,656 (5.5%) operate some form of transportation; 69,177 (11%) a service-related business; and 75,559 (12%) operate other types of small-scale enterprises (CBS 2021).

Nepal is among the least developed countries in the world, with a Human Development Index ranking 143rd out of 189 nations (UNDP 2022). GDP per capita 2017 was $1,223, with 8.2% of the population living under the income poverty line ($2.15 a day).[1] Agriculture is the biggest employer, providing a livelihood for almost two-thirds of the population but accounting for only about 24.1% of GDP (NRB 2024).

Governance and Disaster Management

In 2015, Nepal radically altered its constitution to create a three-tiered federal structure from an earlier unitary structure in an attempt to decentralize government and services. In the three-tier federal structure, each federal-, provincial-, and local-level governance unit is expected to function autonomously. The federal government is responsible for matters of national importance, such as defense, foreign affairs, monetary policy, laws on regulations of insurance, securities, and cooperatives, and health policies. Provincial power includes the management of police, agriculture, industry, trade, cultural preservation/promotions, and land management, among other things. The local or municipal government is responsible for managing local services, basic sanitation, health, education, village/municipal/district assemblies, local courts, and local tax collection, among other things. The three levels of government are concurrently responsible for services such as agriculture, health, and disaster management.

---

[1] World Bank. 2021. Data. https://data.worldbank.org/country/nepal (accessed 19 July 2023).

This structure aims to promote regional development and ensure the equitable distribution of resources and opportunities across the country. The major revenue sources, such as customs duty, excise duty, value-added tax, and income tax, are under the jurisdiction of the federal government (The Kathmandu Post 2015). The provinces' primary revenue sources are house and land registration fees, vehicle tax, tourism income tax, agri-income tax, and local fines and penalties. The local or municipal government's sources of revenue include property tax, house rent tax, land registration tax, and entertainment tax, among other sources.

Nepal's Disaster Risk Management and Reduction Act 2017 legally institutionalized the disaster management system and assigns disaster risk management as a collective responsibility to all tiers of government. Much of the responsibility of planning and governance lies with the local municipal governments. The act also provides disaster management funding at the federal, provincial, district, and local levels.

The National Disaster Risk Reduction and Management Authority oversees the National Disaster Risk Reduction and Management Centre at the Ministry of Home Affairs and operates as the implementing arm of government in disaster risk management. Many other agencies are also involved in disaster risk management and reduction.

(i)   The Office of the Prime Minister and the Council of Ministers coordinate, direct, and facilitate the preparation of national policy and strategy to reduce disaster risk.

(ii)   The National Planning Commission is mainstreaming disaster risk reduction into long-term, periodic, and annual development plans, as well as developing multisector disaster risk management guidelines.

(iii)   The Water and Energy Commission makes recommendations for disaster risk reduction by identifying disaster-prone rivers and streams.

(iv)   The Department of Hydrology and Meteorology generates and disseminates hydrological and meteorological information for water resources, agriculture, and energy. It issues hydrological and meteorological forecasts, and develops operational flood forecasting and early warning systems for major flood prone rivers.

(v)   The Ministry of Federal Affairs and Local Development raises the technical and functional capacity of local authorities on disaster risk reduction and facilitation, ensuring that disaster risk reduction is part of local development plans and that authorities have issued a model local law for disaster risk management.

(vi)   The Department of Mines and Geology operates the National Seismological Centre, which monitors seismological shocks throughout the country through its network of 21 short-period seismic stations and seven accelerometer stations.

(vii)   The Ministry of Irrigation works to minimize future disaster risk through the appropriate design and construction of irrigation schemes and the formulation of policy on water-induced disaster management, flood management, and river training.

(viii) The Ministry of Education raises awareness of disaster risk reduction through programs for teachers, students, and school management.

Post-Disaster Fiscal Risk for Municipalities

The 2015 devolution of power and resources to local governments empowered local governments to plan and execute disaster risk reduction and climate change adaptation plans. This ground-up approach has the potential to be extremely effective in building resilience because people most impacted by disasters now have a say in setting the policies.

The Government of Nepal retains nearly all public sector disaster risk and relies predominantly on *ex-post* financing arrangements. The Prime Minister Disaster Relief Fund is the primary source for immediate relief and rehabilitation of victims. After major disasters, the government often launches an international appeal for donations to the fund.[2] Currently, few *ex-ante* disaster risk financing instruments are in place. Insurance for public assets has been purchased only for a few strategic assets. Post-disaster budget reallocations often mean taking funding away from the existing projects.

On the other hand, municipalities also have limited funds to support disaster relief post disaster. Provincial and local governments are required to allocate 5% of their budgets to local disaster management fund. Rural municipality disaster funds are mostly NRs50,000 and NRs1 million. Similarly, urban municipalities have allocated NRs1–2 million; and metropolitan and sub-metropolitan jurisdictions have allocated NRs2.5–5.0 million (Bhandari 2020). Municipalities can seek more funds from the central government, but money received from the Prime Minister's Relief Fund[3] or international sources must be used only for its intended purpose. The other challenge is that relief funds often do not reach municipalities quickly enough.

---

[2]   NRs239 billion was received in response to the 2015 earthquake and NRs304 million for the 2017 floods.

[3]   Prime Minister's Relief Fund is a special fund established to get the public support during disasters and does not get budgetary support. https://pmnrf.gov.in/en/#:~:text=Contact%20Us-,About%20PMNRF,Trusts%2C%20Companies%20 and%20Institutions%20etc.

2

Agriculture Insurance Penetration—Barriers and Options

Agriculture contributes 24.1% of GDP.[4] Of the country's total land area, only about 21% (3.091 million hectares) can be cultivated,[5] but cultivation actually takes place in only about 75% of the cultivable land. Growth in agriculture was 3.0% in fiscal year (FY) 2024 after rising by 2.8% in FY2023, mostly driven by increased paddy output. Out of total agriculture production, paddy contributes 17% (ADB 2024).

Nepal's history of agriculture insurance goes back several decades. The programs were mostly limited to livestock, although there were a few sporadic pilots on crop insurance. In 2013, the government introduced the Crop and Livestock Insurance Directive (CLID). The insurance authority issued different insurance policies to cover cash crops (tea, ginger, banana, kiwi, etc.); fishery policy; and paddy coverage to promote crop and livestock insurance among poor farmers. It also introduced premium subsidy support. Subsidy support was 50% of insurance premiums in FY2013/14. This was later increased to 75% of insurance premium in FY2014/15 and 80% in FY2022/23. The insurance regulator has designated specific districts to each insurance company as focal points for issuing insurance policies.

Crop insurance policies approved and/or designed by the insurance regulator cover losses due to fire; wind; water (rainfall, unseasonal rainfall, hailstorms, flood and inundation); weather-related risk (drought); land risk (earthquake, landslide); thunderstorm; loss due to crop disease or insect damage to crops; and damage by animals. In addition, these policies cover accidental loss of life up to NRs200,000.[6] Food crop (rice, maize, wheat, barley, and millet) insurance covers loss of yield due to covered risk. The premium rate for rice, maize, and wheat crops is 5%. Table 1 presents subsidy rates for crops, livestock, poultry, and fisheries. The insurance policy only indemnifies crop production losses to the point when the (estimated) revenue for the season is lower than the cost of production.

---

[4]   Nepal Rastra Bank. Current Macroeconomic and Financial-Situation. https://www.nrb.org.np/contents/uploads/2024/08/Current-Macroeconomic-and-Financial-Situation-Nepali-Based-on-Annual-data-of-2080.81-2.pdf.

[5]   Statistical Information on Nepalese Agriculture Sector 2020/21. https://moald.gov.np/wp-content/uploads/2022/07/STATISTICAL-INFORMATION-ON-NEPALESE-AGRICULTURE-2077-78.pdf.

[6]   Nepal Insurance Authority Crop Insurance Policy 2079 Directive. https://nia.gov.np/law/policy.

Table 1: Premium and Subsidy Rates for Crop and Livestock Insurance

Insured	Premium (% of total value)	Premium Subsidies Fiscal Year 2022/23 (% total premium)	Max-Compensation (% of sum insured)
Crops	5%	80%	90%
Livestock	5%	80%	90% (mortality), 50% (morbidity)
Poultry	5% (domestic), 6% (commercial)	80%	90% (mortality), 50% (morbidity)
Fisheries	2%	80%	90%

max = maximum.

Source: Nepal Insurance Authority.

The insurance policy's sum insured is limited to indemnity-based insurance which covers for the cost of input or in some cases cost of cultivation (Praseed Thapa 2018).

Insurance companies are permitted to design new products in addition to those mandated by the regulator. However, interest from insurers has been limited. While some pilot projects, particularly in paddy insurance, are being implemented, they have not been scaled up. The primary drawback of an indemnity-based loss of yield policy is that, in the event of product losses, to assess the losses, crop yield must be estimated field by field.

In addition to indemnity crop insurance policy, the insurance regulator has approved weather index-based crop insurance and area yield index insurance crop insurance.

Challenges to Scaling Up Crop Insurance in Nepal

Agriculture insurance penetration is very poor in Nepal, despite subsidies of up to 80% of the premium. The Nepal Insurance Authority has made it mandatory for nonlife insurance companies to have at least 5% of the total portfolio for crop and agriculture insurance, and crop and livestock insurance increased during the period. It should be highlighted that crop insurance (Table 2) penetration was much lower when compared to livestock insurance (Table 3).

Table 2: Crop Insurance Policy Sold in Nepal

Fiscal Year	Number of Policies	Sum Insured (million NRs)	Total Premium (million NRs)	Subsidy (million NRs)	Claim Amount (million NRs)
2022–2023	7,261	3,573.62	204.41	156.59	126.55
2021–2022	5,206	2,911.85	153.85	115.39	104.05
2020–2021	3,811	1,607.03	82.92	62.19	60.00
2019–2020	4,278	3,003.26	101.04	75.78	86.80

Source: Nepal Insurance Authority Agriculture and Livestock Department.

Table 3: Livestock Insurance Policy Sold in Nepal

Fiscal Year	Number of Policies	Sum Insurance (million NRs)	Premium (million NRs)	Subsidy (million NRs)	Claim Amount (million NRs)
2022–2023	185,940	46,876.07	1,990.84	1,501.011	1,344.67
2021–2022	188,550	40,391.15	1,786.01	1,339.51	545.26
2020–2021	125,190	24,773.19	1,098.78	824.08	562.04
2019–2020	112,580	19,091.61	903.72	677.79	480.97

Source: Nepal Insurance Authority Agriculture and Livestock Department.

Crop insurance penetration is deeper when insurance is linked with crop loans/credits, but penetration of agriculture credit is quite low in Nepal, with only 7.6% of total credit disbursed solely to agriculture (Pandey 2022). The actual disbursement to marginal farmers is much lower, due to reasons such as (i) lack of access to formal banking, (ii) lack of knowledge about various financial products, (iii) poor banking outreach due to high operational cost, and (iv) lack of assets for collateral. Discussions with farmers and the Department of Agriculture indicated that most of the agriculture loans are disbursed to large farmers or to agriculture enterprises. Marginal, small, and medium-sized farmers are rarely able to access the (subsidized) crop loans.

Insurance companies might be reluctant to push crop insurance because crop insurance premiums for most crops are capped at 5%–7% of the sum insured. This might be too low, when considering the risk associated with agriculture in the country. Farming in Nepal is risky due to multiple factors, such as lack of irrigation facilities and access to markets (Thapa 2021). Furthermore, the insurance industry claims that the government does not release its share of premium subsidies on time; and sometimes the delay can be years. According to a statement released by the Nepal Insurance Association in early June 2023,[7] the Government of Nepal had not paid premium subsidies worth NRs3 billion (Dhakal 2023). The report also claims that insurers are unwilling to sell or issue agriculture insurance policies if the subsidy amount is not released.

In Nepal, access to financial services is key to building farmers' resilience. Yet, due to a host of barriers—banking fees, distance from banks, low literacy levels, and cultural barriers that make farmers hesitant to approach formal banks—only 45% of Nepalis use formal banks. Instead, much of the mostly rural population either store cash at home or take credit from the informal sector, where interest rates can soar as high as 48% (Winrock International 2021).

During the focus group discussion, several farmers singled out input cost as the basis of indemnity for crop insurance, and lack of transparency in arriving at the sum insured (input cost for crops and market value for livestock) could also be a major cause of discontent. With no objective criteria for determining the sum insured, local agriculture or livestock technicians decide the sum insured based on their own understanding of the market and other economic factors.

---

[7]   The Nepal Insurers' Association is the umbrella association of nonlife insurance companies.

Box 1

What's Required for Crop Insurance Claim?

- Written claim intimation letter
- Fully filled claim form
- Recommendation letter from concerned technician
- Recommendation from local authority
- Witness, Sarjamin Muchulka
- Recommendation letter from related department office
- Photographs of loss object and/or area
- Cause and estimate of loss with a recommendation letter from the concerned authority
- Procedure on activities applied to check the immediate loss
- Report of concerned technician with detail of loss measurement and amount
- Detail of input cost up to the date of loss

Source: Himalayan Everest Insurance Limited.

An indemnity-based insurance claim settlement requires a list of documents to be submitted for validation of claim (Box 1). Fulfilling such requirements is a tedious hassle for farmers.

Lack of awareness about crop insurance among farmers is another primary reason for low uptake of crop insurance. Farmers in Nepal rarely have information about crop insurance, as the government makes little effort to create awareness (New Business Age 2021, The Kathmandu Post 2018). Notably, Nepali farmers were aware of India's heavily promoted crop insurance scheme (Pradhan Mantri Fasal Bima Yojana), as observed by this report's authors.

The enrollment and claims process for crop and livestock insurance requires complex paperwork, documentation, and compliance with requirements that deter farmer participation, especially those with limited literacy or access to resources. According to the Nepal Agriculture Economics Society (2021), farmers have to visit two to three times to buy a crop insurance policy, although renewal might be possible with just one visit. Loss adjustments are more challenging because farmers depend heavily on agents or technicians to prepare necessary claims documents and loss reports.

According to insurance sector leaders, crop insurance in Nepal faces extreme operational challenges. This is due to difficult terrain, lack of human resources (especially technicians for valuation during enrollment, and surveyors for loss assessment), and procedural difficulties in enrollment and claims adjustments. Insurance executives in Nepal stressed that crop insurance should be made as operationally simple as livestock insurance for it to scale up, a view starkly opposed to that of insurance companies in India.

This report believes that insurance companies might be reluctant to push crop insurance because crop insurance premiums for most crops are capped at 5%–7% of the sum insured (i.e., the cost of input). The insurance premium might be too low as farming in Nepal is risky due to multiple factors, such as lack of irrigation facilities, advisory services, and access to markets (Thapa 2021).

Furthermore, the insurance industry claims that the government does not release its share of premium subsidies on time; and sometimes the delay can be years. This makes it extremely difficult for the insurance companies to settle claims and distribute for next season. According to a statement released in early June 2023 by the Nepal Insurance Association, the Government of Nepal had not paid premium subsidies worth NRs3 billion (Dhakal 2023) due to which insurers are unwilling to sell or issue agriculture insurance policies if the subsidy amount is not released.

> If Nepal's crop insurance program is scaled up in its present form, the premium subsidy burden for the Government of Nepal can quickly become overwhelming. Nepal has been facing a tough fiscal situation due to fall in tourism and the cost of rebuilding post 2015's devastating earthquake, with the reduced tax revenue in 2023 and the government has barely maintained its expenditure from its revenue, it would be difficult in longer days for sustainable subsidy flow from the government.

A number of farmers have singled out input cost as the basis of indemnity for crop insurance as a major cause of dissatisfaction. Apart from the limited indemnity offered by agriculture insurance products, a lack of transparency in arriving at the sum insured (input cost for crops and market value for livestock) could also be a major cause of discontent. With no objective criteria for determining the sum insured, local agriculture or livestock technicians decide the sum insured based on their own understanding of the market and other economic factors. This can be arbitrary and thus a major cause of farmers' dissatisfaction.

If Nepal's crop and livestock insurance program is scaled up in its present form, the premium subsidy burden for the Government of Nepal could quickly become overwhelming. Nepal has been facing a tough fiscal situation due to the decline in tourism and the cost of rebuilding after 2015's devastating earthquake. And with reduced tax revenue in 2023, the government has barely maintained its expenditures. Sustaining the flow of premium subsidies could be difficult, and there have been long-overdue subsidy premium payments to insurance companies and instances of insurance companies stopping the selling of insurance policies due to delayed payments.

Barriers to Scaling Up Crop Insurance

In the past, when only indemnity-based crop insurance programs existed, they performed poorly across the globe. India's Comprehensive Crop Insurance Scheme (1985–1999) had a claims ratio of 575%. Similarly, Bangladesh's indemnity-based crop insurance scheme had a claims ratio of 450%. In the given scenario, even if the Government of Nepal invested in improving systems, it would be difficult to scale-up indemnity-based crop insurance with the current technology and practices.

The administration costs (for enrollment and loss adjustment) are high when compared to the risk premium, as farm holdings are small in Nepal, and thus the premium per farm is not enough to profitably cover the cost of enrollment and of loss adjustment.

Insurance companies stated that technical personnel for both enrollment and loss adjustments is lacking. This barrier is especially difficult to surmount in the short to medium term, which makes it difficult to scale up crop and livestock insurance.

Indemnity based crop insurance is subject to high moral hazard and adverse selection. This is due to asymmetric information between the insurer and the insured, that is, moral hazard problems occur because the insured can take actions which affect the probability of losses and cannot be observed by the insurer. On the other hand, the program sees adverse selection when insured farmers who experience fewer losses choose to opt out of the program, leaving only farmers with a higher probability of losses.

Options for Strengthening the Crop and Livestock Insurance Directive

This section proposes a set of recommendations for strengthening and improving the Crop and Livestock Insurance Directive issued by the Nepal Insurance Regulator. These recommendations are made on the basis of preliminary institutional and operational review for the agricultural insurance program in the country. The recommendations draw on international practices and experiences, including the authors'.

The Government of Nepal should consider a layered subsidy approach toward crop insurance in which the subsidy amount would be optimally utilized based on government social protection goals, without overwhelming state finances. A well-accepted, layered approach is where the government pays or provides a standard level of insurance to all, with the farmer having the option to purchase additional protection as a top-up.

> Approach 1: The government sponsors catastrophe insurance for all farms, and the farmers have the option to top-up by purchasing additional insurance for medium-level risks.

> Approach 2: All farmers are insured at government cost for a fixed amount—two acres,[8] say—and the farmers have the option to purchase additional insurance (for more farms) if required.

The basis of indemnity should be reconsidered to make the product more attractive to farmers. It can be determined based on (pre-agreed) cost of production (per unit) and average yield units. When crop revenue (or approximation) falls short of the investment, the shortfall is indemnified.

Premium rates should be redetermined on an actuarial basis for each district–season–crop combination given the difference in risk exposures. The premium rates were set in 2013 and the crop and livestock insurance scheme has been operating at a loss ratio of 200%. The re-rating can be overwhelming as yield and risk data need to be either collated or generated, and this task should thus be undertaken in a phased manner (maybe starting with paddy).

Reevaluate premium subsidies after analyzing the farmer's willingness to pay (premium). Based on this report's discussions with farmer groups and officials, willingness seems to be higher than estimated to pay for crop insurance. This corroborates Budhathoki (2019), in which about 84% of farmers were interested in purchasing area-based crop yield insurance and were, on average, willing to pay a premium of $42.42/hectare (ha)/cropping season for paddy rice, and $29.52/ha/season for wheat. This amounted to more than three times the price of paddy rice premiums ($9.96/ha/season) and wheat premiums ($8.59/ha/season) under the current subsidized scheme. Another study (Guo 2016) indicates that farmers were willing to pay 3% of household income as a premium for weather index-based insurance for rice when they were aware of the impact of climate change.

---

[8] This number can be changed based on the social protection goals of the Government of Nepal.

Reduce the enrollment and compliance burden. The cost of servicing small and marginal farmers through indemnity-based insurance programs is high for insurance companies due to extensive documentation requirements. Additionally, recovering the expenses associated with distribution and loss assessment is challenging, given the low premiums per farm. Farmers also face difficulties as the process demands numerous documents during both policy issuance and claim settlement. These barriers can be mitigated by adopting area yield index insurance or weather index-based insurance for the state-sponsored portion of crop insurance, particularly for catastrophic risks. For farmers seeking additional coverage, indemnity-based insurance could serve as a complementary option. The subsequent sections examine the feasibility of area yield index insurance and weather index-based crop insurance for Nepal.

Suitability of Area Yield Index Insurance for Nepal

In area yield index-based crop insurance, losses are not assessed based on the yield (individual) of insured farms. Instead, the indemnity is based on the realized (harvested) average yield of an insured area (with homogenous cropping conditions). The insured yield is established as a percentage of the average yield for the area. An indemnity is paid if the realized average yield for the area is less than the insured yield, regardless of the actual yield on a policyholder's farm. This type of index insurance requires historical area yield data on which the normal average yield and insured yield can be established. The average yield for the insured area is determined based on yield sampling exercises that are often called crop-cutting experiments.

While area yield index insurance would be more feasible to scale than Nepal's current indemnity-based crop insurance scheme, it still faces several challenges. High-quality granular yield data is important for the optimal pricing of area yield index insurance, as the insurance premium is determined based on historical crop yield data. Such data for Nepal are only available at the district level. Discussion with field level staff from the Department of Agriculture indicates that only a few crop yield sampling exercises are conducted every season and the yield data may not represent the actual yield, especially in districts with varied topography and a microclimate.

The primary challenge for area yield index insurance would be the large number of crop-cutting experiments required for loss assessment once the program scales up. Currently, the Department of Agriculture deploys a limited number of personnel with numerous duties, including monitoring of various schemes. It would be impossible for ground staff to conduct or oversee the necessary count of crop-cutting experiments. Despite India's long history of area yield index insurance, for example, that country often struggles to conduct the required number of crop cutting experiments every year.

Enrollment for non-loan-linked area yield index-based crop insurance enrollment can become extremely intense operationally, as correct land records are essential to prevent fraud and over or under insurance.

Area yield index-based crop insurance is also subject to adverse selection, and cases of political interference during yield estimation have been reported.

Suitability of Weather Index-Based Crop Insurance

In weather index-based crop insurance, the indemnity is calculated based on deviations of a specific weather parameter measured over a pre-specified period. The contract is structured based on the relation between crop yield losses and weather deviations. An indemnity is paid whenever the realized value of the index exceeds or falls short of a pre-specified threshold.

Weather index-based crop insurance has a number of advantages over indemnity-based crop insurance or area yield index-based crop insurance, especially when operating in data-sparse conditions. Weather index-based crop insurance does not suffer from moral hazard and adverse selection. The payout/loss calculation is more transparent and helps build trust (in insurance). It is operationally less intense, as loss assessment is not based on crop-cutting experiments.

The primary challenge of weather index-based crop insurance or index/parametric insurance is basis risk, which is the difference between loss experienced and payoff from parametric insurance. Basis risk arises because of (i) the difference between weather recorded for payoff calculation and actual weather at the insured location, and (ii) due to poor correlation between the loss index and actual losses. The former is described as spatial basis risk, and the latter is termed model basis risk.

Reducing spatial basis risk in weather index-based crop insurance needs a dense network of weather observation stations. The Department of Hydrology and Meteorology maintains a network of weather stations but unfortunately this network is not dense enough. The insurers thus needs to invest in weather monitoring infrastructure. It is a significant investment in equipment and maintenance cost. In addition, third-party data auditors are required to ensure transparency. Appendix 4 details automated weather stations.

Similarly, to reduce model basis risk, the relation between weather parameters and crop losses/yield reduction should be well established. But this can be challenging as historical granular yield data or loss data might not be available, which is necessary to establish the relationship between yield losses and underlying weather parameters.

Basis risk can be especially high when the relation between weather parameters and crop losses/yield reduction are not well established. Moreover, historical yield data might not be available, which is necessary to establish the relationship between yield losses and underlying weather parameters. Weather index-based crop insurance products can be especially difficult to explain to farmers and the sponsoring governments.

3

Status of Disaster Insurance in Nepal

Insurance is accepted as a critical component for recovery from disasters. Relief from aid agencies and/or government is often too little or too late to rebuild, and, similar to crop insurance, the penetration of property insurance and home insurance is abysmal. While a detailed study on insurance penetration for property, households, and SMEs is beyond the limited scope of this study, reports such as Investopaper (2023) indicate that insurance penetration for nonlife insurance was 0.65% of the gross premium to GDP in 2021–2022. Post-disaster insurance claims data from the 2015 earthquake and 2017 flood also indicate extremely low insurance penetration.

Losses due to the 2015 Gorkha earthquake: The estimated economic loss from the 2015 Gorkha earthquake was $5.1 billion, while insured losses were only $175 million. No part of the world is more vulnerable to natural catastrophes than Asia, where just 12% of economic losses—$152 billion of $1.24 trillion—were covered by insurance. This means that, in most major disaster events, virtually all losses are uninsured and local populations are entirely dependent on financial support from the federal government or international aid agencies for recovery (Trendafiloski 2020).

While insurers in Nepal offer property insurance[9] policies (insurance against perils such as earthquakes, floods, landslides, fire, windstorms, hail, water related damage, thunderstorms, etc.) and loss of revenue policies (a consequential loss policy),[10] the policies are often purchased only by large business houses. SMEs and homeowners either do not purchase property insurance policies or access to them is not easy.

---

[9] See Nepal Insurance Company. Property Insurance. https://en.nepalinsurance.com/policy/property-insurance.

[10] It seems loss of revenue policies can be purchased only if the insured already has a property or mechanical breakdown insurance policy. To ensure maximum benefit from a loss of revenue/profit cover, the insured have to diligently maintain books of account.

SMEs are also highly vulnerable to disasters due to their insufficient financial and technical capacity. They have limited awareness of climate risks because of low human resources and lack of information. Unlike large enterprises, SMEs often operate in smaller local geographies, which increases their risk concentration. Australia and Japan have the highest premium per capita to GDP ratio (Figure 2).

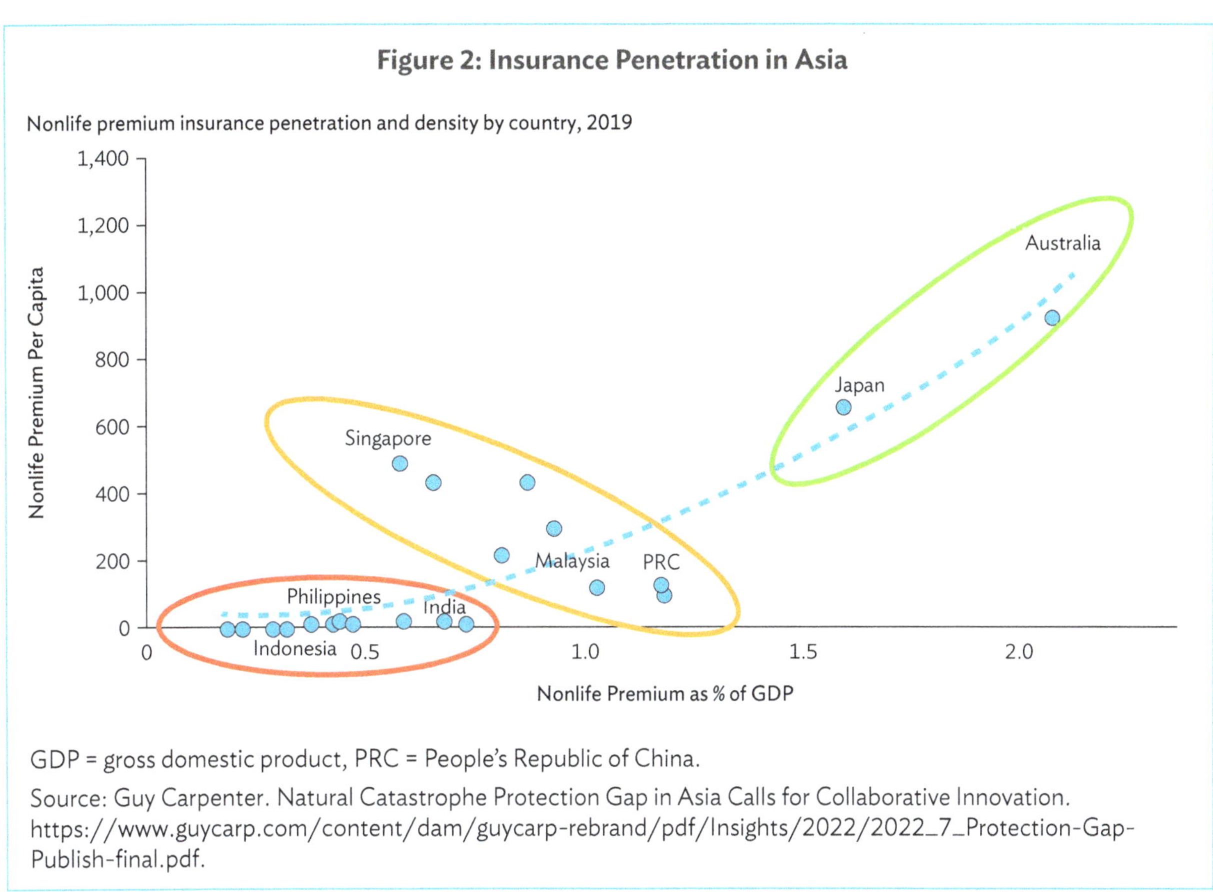

Figure 2: Insurance Penetration in Asia

GDP = gross domestic product, PRC = People's Republic of China.

Source: Guy Carpenter. Natural Catastrophe Protection Gap in Asia Calls for Collaborative Innovation. https://www.guycarp.com/content/dam/guycarp-rebrand/pdf/Insights/2022/2022_7_Protection-Gap-Publish-final.pdf.

Farmers and SMEs are profoundly affected by disasters such as agricultural failure, floods, landslides, and earthquake. Insurance solutions have failed to reach the most vulnerable, despite premium subsidies and other efforts by the government and civil society organizations.

Barriers to Scaling Up Disaster Insurance

Lack of awareness is a major barrier to purchase of property insurance. In India, a survey in 2015 by Bajaj Allianz General Insurance Company found that citizens often assume that home insurance is expensive (Bajaj Allianz General Insurance 2015).[11]

---

[11]   M. Mohanka. 2023. The implications of Uniform Civil Code for taxation and inheritance. *Mint.* 11 July. https://www.livemint.com/money/personal-finance/governments-proposed-uniform-civil-code-in-india-implications-for-taxation-and-inheritance-laws-11689097517664.html.

Nonlife insurance companies received claims worth NRs2.5 billion. The highest claims were received from industries for flood damage to warehoused goods, machinery, and physical infrastructure. Insurance claims for damage to private cars and small public vehicles were also significant. Claims for losses to livestock and crops was only NRs23 million. Claims for crop losses were much lower compared to the estimated damage, as farmers are reluctant to buy agricultural insurance (The Kathmandu Post 2017).

Inadequate distribution channels is another barrier, as insurers do not have an adequate number of branch offices. Moreover, due to low demand, insurance agents lack awareness of all the features of property insurance.

Lack of data to accurately price risks is a major challenge for Nepal because of poor data collection, outdated risk information, and access to risk data. This makes it difficult to both price the risk accurately and attract international reinsurance.

Capacity gaps for data analysis, product development, and underwriting are a challenge for a small developing nation like Nepal. The capacity gaps make it especially difficult to research and develop microinsurance products for disasters.

Parametric Insurance for Crops

4

Parametric (or index-based) insurance is an insurance policy that indemnifies against the intensity of a predefined event instead of indemnifying actual losses. This type of insurance can protect against the financial impact of events that are difficult to predict or quantify using traditional insurance methods. This can be particularly valuable in situations where rapid response is needed, such as in the aftermath of a disaster or other catastrophic event. Parametric insurance is covered in greater detail in Appendix 1.

Index insurance has multiple advantages and can provide faster payouts than traditional insurance policies. Index insurance policies are simpler to administer and reinsure, as they rely on observable and verifiable data. This makes them particularly suitable in contexts where validated loss data are available. Periodic updates to historical hazard data further enhance the claim settlement process for index-based policies. In a country like Nepal, index insurance products are easier to deploy and may have a lower cost of administration than traditional indemnity-based insurance, especially the cost of loss adjustments. Index insurance also has less scope for moral hazard and adverse selection.

Indemnity-based insurance products provide coverage for class 1 houses, such as concrete houses, while excluding mud mortar houses due to their higher risk profile. Insuring mud mortar houses poses a challenge because of the elevated risk, resulting in higher premiums, which are often unaffordable for low-income individuals (Box 2).

Index insurance has two disadvantages: (i) basis risk,[12] which makes it difficult to cover low severity risks, and (ii) difficulty in product design in a sparse environment. Parametric insurance product concepts that seem suitable for the surveyed municipalities include:

Drought index-based insurance:[13] Drought is a primary hazard in Nepal, mainly caused by uneven and irregular low-monsoon rainfall. The lack of irrigation facilities exacerbates the effect.

---

[12] Basis risk is the difference between payout and losses.

[13] Many drought indexes seem suitable for Nepal, such as the Standard Precipitation Index or the Palmer Drought Severity Index, or the Standardized Precipitation Evapotranspiration Index, which is a meteorological drought index that uses both precipitation and potential evapotranspiration data to assess drought conditions at different time scales.

Box 2

Can a House of Mortar Be Insured?

In the 2015 earthquake, 95% of the houses that collapsed were built of low-strength unreinforced masonry, often with mud mortar. While Nepal has standardized building code, it lacks building inspectors, while an absence of local government mechanisms and an overburdened judicial system are also factors undermining enforcement. The lack of human and technical capacity and materials is also hampering implementation. As close to 80% of all houses are owner-constructed, compliance with the building code strongly depends on owners' understanding of the risks, costs, and benefits of following the Nepal National Building Code. After the earthquake, implementation of building code in urban areas is mandatory.

Sources: Ministry of Urban Development. Nepal National Building Code. https://www.moud.gov.np/pages/nepal-national-building-code; World Economic Forum (2015).

In the last 50 years, Nepal has witnessed five severe droughts (1972, 1979, 1994, 2005, and 2015) and several medium to minor ones. Parametric insurance product concepts that can be explored for the Terai region are:

- Drought insurance for paddy: Paddy is water dependent crop and, if rainfall is not adequate, it can lead to yield losses.

- Sowing failure insurance: Delay in the monsoon can lead to sowing failure for crops other than paddy.

Fodder deficit insurance:[14] Cost of fodder jumps after drought due to lack of biomass production. This raises the cost of the livestock maintenance.

Flood index-based insurance:[15] Flood is probably the most damaging. Historical data show major floods in the Tinao Basin (1978), Koshi River (1980), Tadi River Basin (1985), Sunkoshi Basin (1987), and a 2017 flood that affected 1.7 million people. In interviews, local government officials and SMEs have expressed strong demand for property insurance against fire and flood damage. Severe floods can submerge local shops, factories, and farms, and wash away livestock, causing damage in millions. A comprehensive explanation for flood- and water-induced risk detail is provided in Appendix 2, Appendix 3, and Appendix 4 for further reference. Parametric insurance products that can be explored for the Terai are:

- Flood insurance for crops: Insurance against crop losses due to flood. Note that the loss pattern differs based on crops.

- Flood insurance for warehouses and micro, small, and medium-sized enterprises (MSMEs): Flood water can enter warehouses and small business. The loss function should be a combination of flood duration and flood extent.

---

[14]   This is an interesting setting, where the insurance can pay before the actual loss, because the effect of drought (i.e., lower fodder availability) is felt next season.

[15]   Flood index insurance products can be difficult to design due the complexity of floods and lack of exposure data.

Income/wage loss for daily wage workers: Daily wagers are among the lowest income earners and live in areas that are most exposed. During floods, their income stops, while expenses go up due to the cost of medicine or higher cost of food and fodder, forcing them to distress-sell their assets. Flood can be used to insure wage losses.[16]

Five hundred industries were shut down at the Morang Industrial Corridor due to Nepal's devastating 2017 flood. Most of these industries raised insurance claims for damage to machinery and warehoused goods. But the (daily wage) workers in these industries did not have a mechanism to compensate them for the loss of wages caused by the shut down or absence from work (Ghimire 2017).

Earthquake index-based insurance: This type of policy pays out a predetermined amount of money if an earthquake of a certain magnitude occurs within a specified area. This can be particularly useful in earthquake-prone areas where traditional insurance may be difficult to obtain or prohibitively expensive. Different government-backed earthquake schemes across the globe are detailed in Appendix 7. Parametric earthquake insurance is implemented in Nepal and attracting interest (Box 3).

Box 3

Parametric Earthquake Insurance Solution in Nepal

The Upper Trishuli-1 Hydropower Project is a 216-megawatt greenfield run-of-river hydropower project. Progress was severely hindered by the 2015 Nepal earthquake, which left traditional insurers reluctant to provide earthquake cover in the location of the dam. This in turn affected investment potential. Swiss Re, Aon, and the International Finance Corporation have structured a parametric insurance cover which enabled the project to move forward. The index is triggered by shaking intensity at the project site and based on United States Geological Survey shake maps after a major quake.

Source: Gallin (2023).

Landslide insurance for mountain regions: Steep slopes, unstable geography, and intense monsoon rains combine to make Nepal extremely hazard prone, and parts are highly vulnerable to landslides. Landslide insurance can be used to insure loss of agriculture land (not just agriculture produce). Appendix 3 discusses precipitation index-based landslide insurance.

Multi-hazard parametric meso-level insurance products can (i) protect the loan portfolio of (agri focused) financial institutions against large scale defaults, or (ii) provide financial support to agri-businesses when large-scale crop losses in their catchment area lead to business interruption.[17] These policies are designed to address the unique needs of groups or organizations that may not have access to traditional insurance products or may face barriers to obtaining coverage. Examples of meso-level insurance are covered in Appendix 2.

---

[16] Read about Oxfam's wage loss insurance at the World Food Programme. https://www.wfp.org/news/casual-labourers-bangladesh-benefit-newly-launched-flood-insurance-scheme.

[17] Meso-level insurance targets risk aggregators, such as local governments, banks, or agri-businesses. Such insurance products are primarily designed to protect the portfolio risk of the policy holder.

5

Sub-Sovereign Insurance Pool

Nepal Catastrophe Insurance Pool for Municipalities

Phase 1 of ADB (2019b) recommended the Government of Nepal consider creating a sub-sovereign Multi-Hazard Catastrophe Insurance Pool for Agriculture and Disaster. The following section builds on the recommendations and defines its contours. Appendix 5 details practices across the globe for such pools.

Conceptually, the Nepal Catastrophe Insurance Pool (NCIP) should be an insurance pool to (i) insure municipal government-owned structures and (ii) insure against governments post-disaster liabilities (for relief and rehabilitation). But if a pool is formed, it would be prudent to expand the scope to bring more businesses or households under insurance to increase the risk capital. Yet, there is no standard framework for establishing a pool, and existing approaches may not be effective under the given conditions. Based on information gathered through field studies, the following framework is proposed (Figure 3).

The municipal government should identify the (uninsured) farmers, low-income households, or MSMEs they want to bring into the insurance safety net. Municipal governments may also choose to invite others (higher-income citizens or businesses) to purchase insurance from the pool at market price. The pool technical team should help municipal governments understand the disaster risk, financial needs, and the frequency and scale of payout for the insurance policy sought by the municipal governments (depending on the type of hazard and hazard frequency that those governments want to insure against).

Once participating municipal governments identify the end client/beneficiaries, the pool technical team's role would be to design the parametric insurance products, take client feedback, set the premium rates, set the guidelines for loss adjustments, and secure the necessary approval from the management board. It should be ensured that there is no cross-subsidization of the premium across municipalities and that the premium is determined based on the risk they face.

The pool should then decide the risk it wants to retain and seek reinsurance for the rest. International reinsurers such as SCOR Re and Swiss Re have expressed interest and others can be explored. The Agriculture Insurance Company Limited of India also intends to expand outside India and can become a strategic partner of the pool; domestic reinsurers are keen.

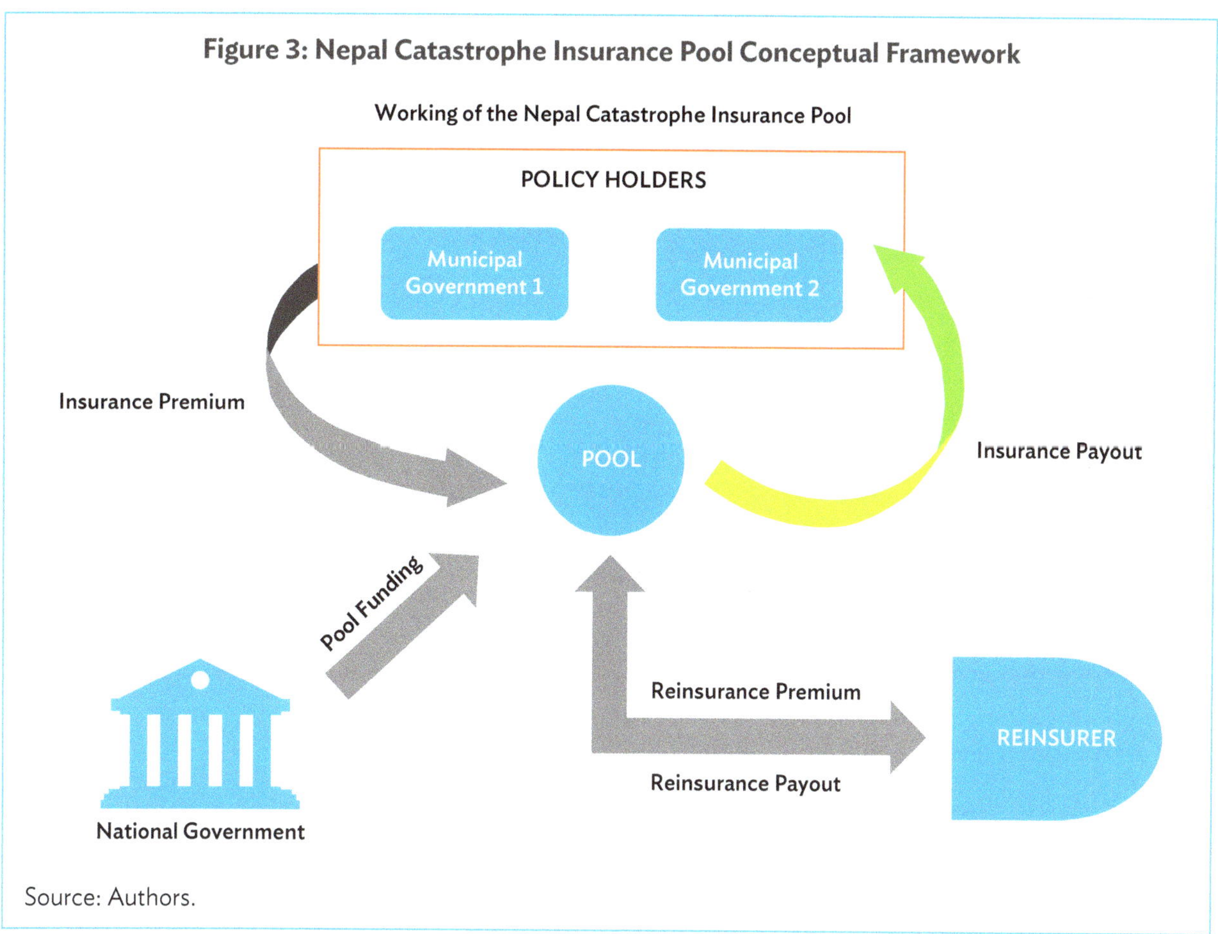

Figure 3: Nepal Catastrophe Insurance Pool Conceptual Framework

Source: Authors.

Municipalities will purchase parametric insurance from the pool. During the policy period, the Pool Technical Team will monitor data, calculate payouts, and channel any payout to the municipalities. The Pool Technical Team will also inform reinsurers of any payment, as per reinsurance policies.

Municipal governments should have two options for insuring their disaster risk reduction liabilities. They can either purchase (i) an insurance policy for themselves or (ii) group insurance policy for end beneficiaries. In the former case, the pool pays compensation (if any) directly to the municipal governments, without considering how they will utilize the money, while in the latter case, the compensation is distributed directly (by the pool) to the individual beneficiaries.

The pool can play an important role in the initial stages identifying and enrolling farmers. A village-level enterprise/nongovernment organization can be enrolled that can help (i) collect information for risk analysis and mid-season monitoring of claims, (ii) support distribution and loss adjustment, and (iii) support grievance handling. The proposed solution envisions the municipality as an aggregator, leveraging its ability to efficiently reach individuals and local MSMEs (Figure 4).

In Nepal, female farmers are often in disadvantaged positions due to societal norms and are not able to take advantage of schemes. This report strongly suggests that women be primarily engaged as village-level entrepreneurs, *Beema Bahini,* which generate employment for women and make it easier for female-led households to join the scheme.

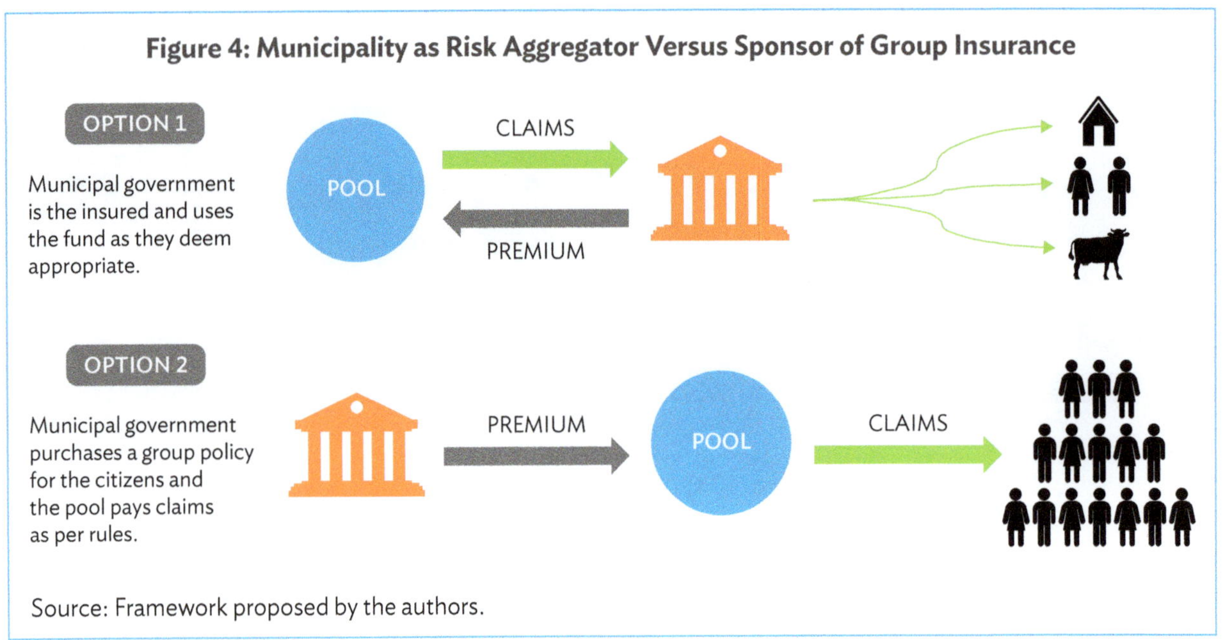

Source: Framework proposed by the authors.

Administrative Structure of Nepal Catastrophe Insurance Pool for Municipalities

The pool should be overseen by a steering committee comprised of members representing (i) central ministries (e.g., agriculture, finance), (ii) participating municipalities, (iii) the insurance regulator, and (iv) funding agencies (who seed the pool). External experts with experience of catastrophe sovereign-risk pools should be invited to the committee for guidance. Figure 5 presents a structure for managing such a disaster insurance pool.

The pool management board should be comprised of domestic representatives of reinsurers from the Ministry of Agriculture and an international insurance consulting firm or brokers (which would be responsible for reinsurance placement, government relationships, and outreach in the initial period).

The pool's technical team should be comprised of (i) actuaries and insurance product design experts; (ii) subject matter specialists from agriculture, hydrometeorology, seismology; (iii) distribution specialists; and (iv) monitoring and evaluation experts. The role of the pool's technical team would be to (i) design and price insurance contracts, (ii) pay commissions to distribution channels, (iii) support the pool management board in placing reinsurance, (iv) distribute claims for parametric insurance, (v) monitor losses, and (vi) identify losses missed by the parametric insurance and devise a fair methodology to support such households.

One of the most important roles of the pool would be to (i) handle farmer-level grievances, and (ii) municipal government-level disputes, especially if losses exceed reinsurance payments. It is important to understand that the NCIP can only be a part of a comprehensive financial protection strategy. The parametric insurance role is to provide rapid liquidity in the immediate aftermath of severe disasters, but the scope will always be limited. Other financial instruments, such as contingency funds and contingent loans, should be used to finance the cost of more frequent disasters. Optimally, the Government of Nepal needs to use different financial instruments to address different needs with different cost implications.

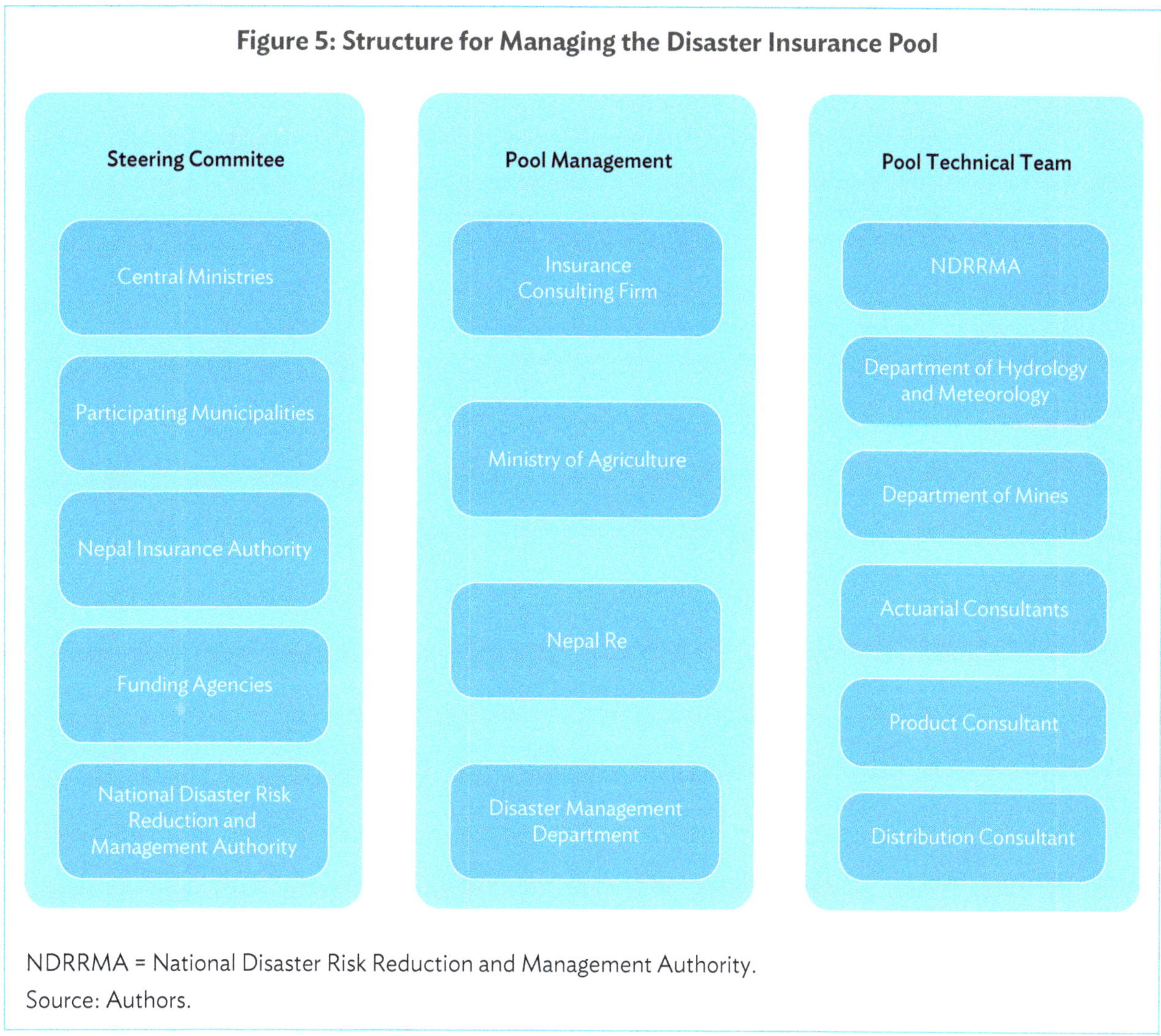

NDRRMA = National Disaster Risk Reduction and Management Authority.
Source: Authors.

Ideally, the NCIP should absorb programs such as crop and livestock insurance, but it might not be possible in the initial phase. The NCIP should therefore work in tandem with existing insurance schemes or other post-disaster financing arrangements, and support distribution of such schemes to create more comprehensive risk protection measures.

An alternative approach would be to keep parametric insurance separate from the Crop and Livestock Insurance Directive (CLID). During enrollment, farmers may choose to purchase additional CLID insurance. Similarly, eligible business houses may choose to purchase property insurance via the pool. The pool should function as a distribution agent (earning a distribution commission) for the CLID or property insurance, or better, get special permission to act as insurer themselves. But the second option will be prudent only if the CLID premium is recalculated actuarially.

Funding Nepal Catastrophe Insurance Pool for Municipalities

The Nepal Catastrophe Insurance Pool needs initial capitalization: The Government of Nepal should secure financing through internal resources or explore funding opportunities from multilateral agencies. Financing participation from domestic insurers, reinsurer banks, and private institutions is crucial to success. Other international donors, such as ADB, the Swiss Agency for Development Corporation, International Finance Corp/World Bank, and KfW, might also be willing to seed the pool. Donors are important for the initial phases, not just for financial support, but also because they provide technical resources needed for the success of the pool. A pool needs funding and participation from different stakeholders, and the pool fund is proposed for claims payout to create the reserves and investments (Figure 6).

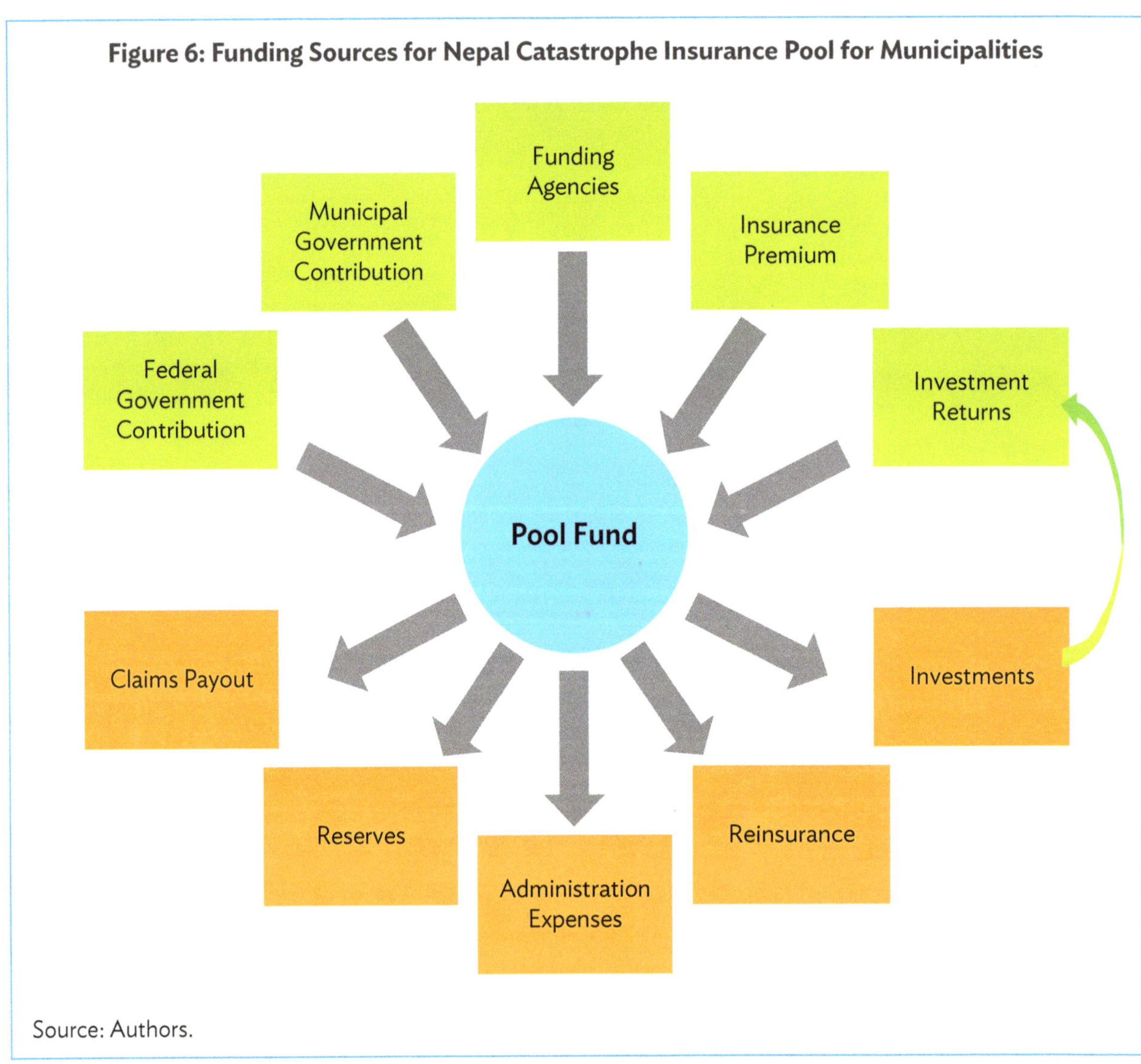

Figure 6: Funding Sources for Nepal Catastrophe Insurance Pool for Municipalities

Source: Authors.

The Government of Nepal and municipal governments that previously relied on donor support may find it challenging to start paying an insurance premium for disaster risks with national resources. Pools require up-front payment of an insurance premium, facilitating a shift toward proactive risk management. Participating municipalities will have to pay an insurance premium upfront that reflects their actual risk exposure in exchange for the insurance coverage, thus making payments predictable installments before disaster strikes.

The NCIP can succeed only with strong political commitment. Strong intent and coordination between the Government of Nepal and municipal governments has to be demonstrated during the initial phases to attract funder and reinsurer interest. Governments should therefore also provide initial capitalization for the NCIP from their disaster risk financing fund.

Establishing and operating a pool requires expertise from various domains to ensure its sustainability and effective functioning, both technically and operationally. It is envisioned that professionals from diverse fields will contribute to the pool's success, either on a full-time or part-time basis (Figure 7).

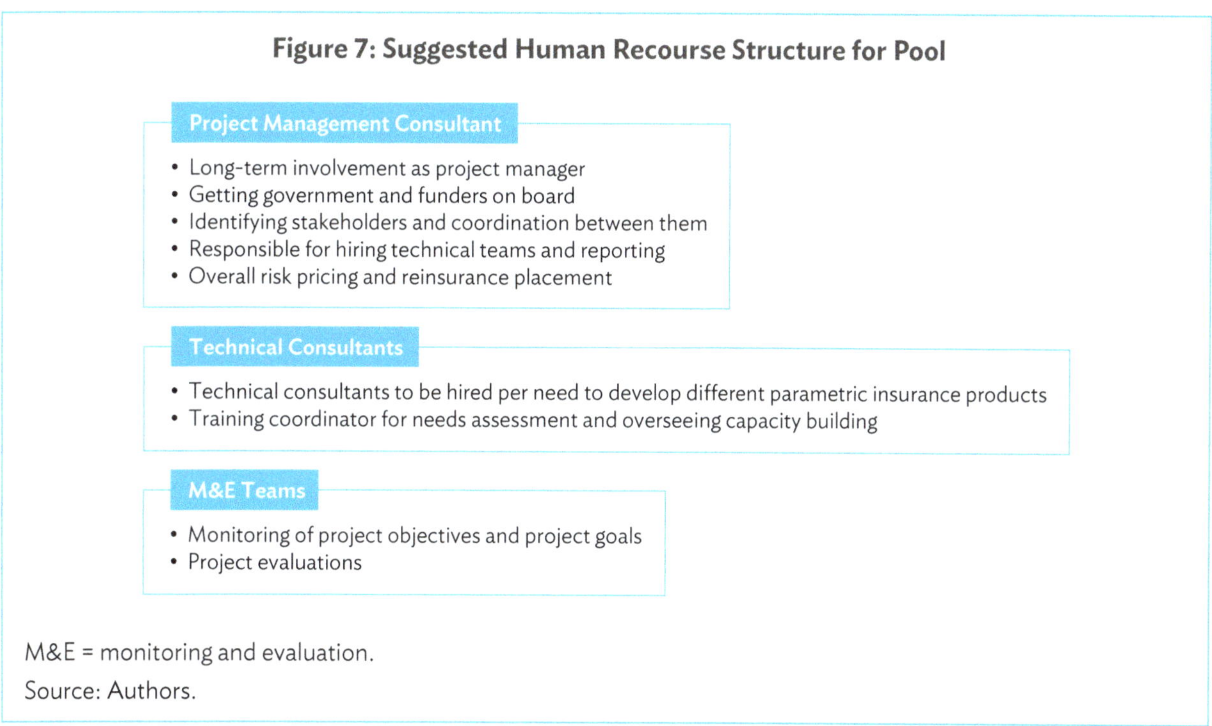

Figure 7: Suggested Human Recourse Structure for Pool

M&E = monitoring and evaluation.
Source: Authors.

NCIP earns premium from (i) disaster insurance premium from municipal governments, (ii) crop insurance premium and subsidies (farmers, municipal governments, and the Government of Nepal), and (iii) premium from SMEs and financial institutions.[18] The pool will also earn a distribution (agent) commission of 10%, which would be ploughed back into the pool. The pool can have layers of insurance coverage and different deductibles for different members. However, the NCIP can be sustainable only with assured premium financing. Allocating budget for the payment of premiums is not generally a permanent part of budgetary processes without a strong central directive.

---

[18]   Financial institutions with exposure to the local economy and local SMEs should be invited to be a part of the pool.

Getting Stakeholders Onboard and Developing Project Strategy

Preliminary discussion with municipal governments, industry, and insurance regulators has indicated strong support for parametric insurance. But before launching the project/pilot, it will be prudent to sign memorandums of understanding with (i) the Ministry of Agriculture and Ministry of Finance for extension of premium subsidy support to the program,[19] (ii) the Department of Meteorology and Hydrology for data sharing arrangements,[20] (iii) the insurance regulator for a sandbox environment for parametric insurance,[21] and (iv) municipal and provincial government (once the pilot locations are confirmed) for pilot approval. Other potential stakeholders invested in Nepal's disaster risk management ecosystem (especially multilateral agencies) should also be identified.

Conducting High-Level Technical and Operational Feasibility Assessment

Design of parametric insurance products needs granular (and historical) crop yield data,[22] exposure data, (insured parameter's) event data, and vulnerability data. While Nepal's data ecosystem is rapidly being upgraded, a deeper study of the data ecosystem is needed[23] (data generation, availability, accuracy, and sharing constraints).[24] Also, a deeper analysis of existing historical data for the proposed pilot locations should be undertaken.[25] This study should identify gaps in historical data and analyze suitable methodologies to plug the gaps.

For the pilot, set up of necessary data monitoring infrastructure might be required. This includes installation of automated weather stations and soil moisture sensors, and building remote sensing technology products for crop, flood, and landslide monitoring (Appendix 4 details automated weather stations). The feasibility analysis should explore the need, financial liability, and operational challenges of erecting and maintaining data monitoring infrastructure.

A pilot can be structured after a critical assessment of needs, clarity in governance, functionality of the insurance pool, and multisector participation and partnership. Management structure should be learned so that it can operate smoothly.

---

[19] The ministry is reluctant to support new pilots with premium subsidies.

[20] The Department of Hydrology and Meteorology has provision to sell data but a blanket sharing arrangement for the project will prevent delays.

[21] While the Nepal Insurance Regulator is extremely supportive of experiments, meso-level insurance is an entirely new concept for the country.

[22] In the opinion of this report, not enough crop-cutting experiments are being conducted to generate an accurate crop yield estimate.

[23] This analysis was beyond the scope of this study.

[24] The Department of Hydrology and Meteorology may find it difficult to share river data for strategic locations or there may be delay in sharing data after an event, which would lead to delay in claim payout.

[25] Data include daily weather data (real and model generated), floods, crop yield, earthquake, landslides, etc.

Funding structure will be defined in the feasibility study. The study should provide clarity on how long a pool will function and whether it will be designed to dissolve based on government and other stakeholder feedback. In addition, the structure suggested should avoid political influence, which can generate partiality.

The insurance regulator committed to support innovations on insurance and pilots, and regulators will require clarity on the framework and supervision of such new innovations.

In addition, the feasibility analysis should define the features of the data management software required to manage the instruments and retrieve and quality check the data.

Due to the lack of historical (real) weather data, actuaries often have to use model generated weather data sets like Climate Hazards Group InfraRed Precipitation with Station Data (CHIRPS), Department of Hydrology and Meteorology Gridded Data, or European Center for Medium-Range Weather Forecasts Reanalysis. The suitability of modeled data sets for insurance in Nepal should be tested, as should satellite imagery for loss assessment.

The feasibility analysis should identify gaps of parametric insurance specifically for Nepal. It should consider factors such as willingness to pay for targeted citizens, government willingness to provide subsidies, and availability of reinsurance.

Building Partnerships

Nepal is a difficult terrain to work in, with diverse risks and not enough data. It would be optimal if multiple stakeholders could come together on a common platform for the pilot and subsequently become part of the proposed disaster pool.

The pilot should explore partnerships with existing experiments and build on the risk analysis already in development. Partnering with pilots, such as (i) the flood insurance project funded by InsuResilience Solutions (Reinsurance News 2021), and the (ii) crop insurance initiatives led by the United States Agency for International Development (USAID)/Tayar Nepal, would allow cross learning and strengthen the ecosystem. Civil society organizations such as Practical Action have been involved in disaster risk mitigation in Nepal, which will bring in institutional knowledge to establish ground operations. Building on existing pilots should defray the cost of data gathering, data analysis, and infrastructure establishment. Potential partners include:

- Technical organizations such as the International Centre for Integrated Mountain Development have generated substantial information on disaster risk in Nepal.

- Government technical agencies such as the Department of Hydrology and Meteorology and the National Disaster Risk Reduction and Management Authority can help ease access to data and optimize insurance products. The National Disaster Risk Reduction and Management Authority's Building Information Platform Against Disaster could be instrumental for accessing real-time event data. The presence of these organizations in the pilot will also help build government confidence.

- Also important is involvement from multilateral organizations such as the USAID, which has been involved via its Feed the Future initiative for more than a decade. Similarly, GIZ, Swiss Agency for Development Corporation, United Nations Capital Development Fund, among others, have invested in building resilience in Nepal.

Developing and Testing Parametric Insurance Products

- Feasibility should be followed by a stakeholder alignment to help national stakeholders better understand potential and the challenges of disaster insurance, and to clearly define success criteria for a pilot.

- Develop and launch parametric insurance products designed specifically for targeted municipalities. While simple weather index insurance and flood index insurance (for crops) should be launched as soon as possible, other parametric insurance such as landslide, earthquake, or flood insurance for MSMEs may need more time to develop.

- The pilot should consider building capacity within the country to research and design parametric insurance products and manage the pool. In addition, strong focus should be on building adequate data management, analytics, and remote sensing capabilities specifically for insurance. Notably, Nepal already has high-level remote sensing technology skills.

- The pilot should also consider building data warehouses, creating mobile applications for field data collection and claims monitoring, and creating artificial intelligence/machine learning tools for automated crop risk and claims alerts.

Test the barriers to timely claims assessment and claims settlement.[26] For scheme success, benefits must reach end beneficiaries quickly post disaster. Lessons from sluggish relief after Nepal's 2015 earthquake are still fresh—the government failed to distribute aid or commence reconstruction even 4 months after the earthquake (MacAskill and Sharma 2015).

- Develop an exit strategy for external consultants and external involvement. The strategy has to be developed in tandem with a capacity-building plan.

Implementation Challenges for the Pilot

Stakeholders' goals are not aligned: Each stakeholder has different expectations and apprehensions. It is essential that major stakeholder expectations are documented and their goals aligned—in government, the insurance sector, and financial institutions (including livelihood-focused microfinance institutions). It is also important to identify people within each important stakeholder group to champion disaster insurance internally.

Lack of comprehensive feasibility studies: Nepal needs a detailed plan for technical, operational, and institutional analysis of the disaster and crop insurance ecosystem. Most earlier studies are extremely limited in scope, lacking either geographic coverage or technical depth. For example, area yield index insurance might be technically feasible but it will be extremely difficult to scale up due to the lack of field technicians.

Capacity gaps: Stakeholders in Nepal are unfamiliar with technical and operational aspects of disaster and crop insurance, given the novelty of concepts such as parametric insurance and use of remote sensing for insurance.[27] Policymakers need deeper appreciation of the scope, technical possibilities, and challenges of disaster insurance products, while the insurance and associated sectors need product design and underwriting capacities.

---

[26] Experiences from India show that claims assessment and claims dispute can take significant time (more than 6 months) due to operational challenges, which causes extreme dissatisfaction among farmers.

[27] A few organizations, such as the International Centre for Integrated Mountain Development and technical governmental agencies, have excellent technical knowledge but lack experience in parametric insurance.

Few effective champions are advocating for disaster insurance in Nepal: Multiple efforts have supported disaster insurance through pilots, studies, and workshops. But without local champions with access to high-level stakeholders to continuously advocate for the cause (regulators, agriculture ministry, finance ministry, and industry), creating momentum is difficult.

Proposed Way Forward for Pilot

Form a disaster insurance alliance/platform within Nepal: Multilateral organizations (ADB, United Nations, GIZ, USAID), technical agencies, research organizations, and nongovernment organizations have put in substantial effort via pilots, research, and advocacy. The lessons have been significant but that has not built the momentum to punch through the barriers. In an alternative approach, organizations would form an alliance and pool their resources and institutional knowledge to achieve their common goals.

Comprehensive feasibility study for Nepal on the crop and disaster insurance ecosystem: This report suggests a detailed study of crop and disaster insurance in Nepal and the feasibility of potential insurance products. This analysis should include (i) risk assessment (agriculture sector, SMEs and MSMEs, government), (ii) technical and operational analysis of insurance schemes sponsored by the Government of Nepal and other disaster insurance schemes, (iii) identifying strategies to scale up (or not) existing schemes, and (iv) identifying opportunities for new disaster and crop insurance products.

Stakeholder alignment: The alignment workshop should be organized for the three levels of government, insurance sector organizations, agriculture-focused financial institutions, disaster management organizations, and other stakeholders. It should include multilateral organizations to develop a better understanding of potential and the challenges of crop insurance.

Pilot: The most effective way to learn about disaster insurance is to develop pilot tests in which insurance contracts are developed, and distributed and loss adjustments and grievance handling mechanisms set up. A framework for establishing a disaster insurance pool is proposed (Figure 8).

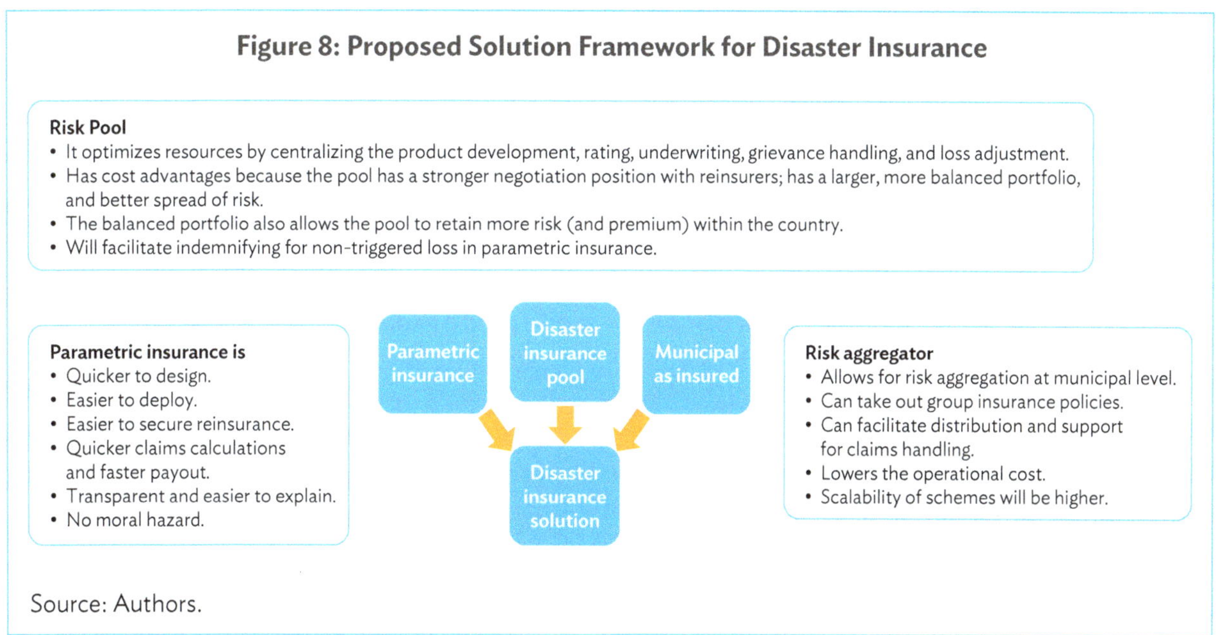

Figure 8: Proposed Solution Framework for Disaster Insurance

Risk Pool
- It optimizes resources by centralizing the product development, rating, underwriting, grievance handling, and loss adjustment.
- Has cost advantages because the pool has a stronger negotiation position with reinsurers; has a larger, more balanced portfolio, and better spread of risk.
- The balanced portfolio also allows the pool to retain more risk (and premium) within the country.
- Will facilitate indemnifying for non-triggered loss in parametric insurance.

Parametric insurance is
- Quicker to design.
- Easier to deploy.
- Easier to secure reinsurance.
- Quicker claims calculations and faster payout.
- Transparent and easier to explain.
- No moral hazard.

Risk aggregator
- Allows for risk aggregation at municipal level.
- Can take out group insurance policies.
- Can facilitate distribution and support for claims handling.
- Lowers the operational cost.
- Scalability of schemes will be higher.

Source: Authors.

6

Recommendations

Carry out weather index-based crop insurance against loss of yield for small and marginal rice farmers. Farmers and local stakeholders say that crop insurance is needed, hence the study strongly recommends. The study also recommends post-disaster assistance to homeowners and small businesses using parametric index insurance products.

Involve municipalities to increase agriculture insurance penetration in Nepal, which now is extremely low. Active municipal participation is important for scalability and sustainability. Municipalities can act as risk aggregators and will make distribution and disbursement easier. A proposed pilot will provide a platform for municipalities to use disaster relief funds more effectively and gain the social commitment of the Government of Nepal. Municipal government participation will also improve communication among stakeholders and can ease the addressing of other operational challenges, such as high administrative costs.

Develop a transparent, well-defined, and acceptable procedure to determine sum insurance. Farmers are extremely dissatisfied with the input-cost basis of the sum insured for crop insurance.

Redetermine premium rates on an actuarial basis, considering the underlying risk exposures. This will help increase insurance company participation and give reasonable comfort to reinsurance companies.

Conduct proper calibration and validation before finalizing the triggers and exits. A serious disadvantage of the parametric insurance approach is the inherent basis risk. High basis risk may create dissatisfaction among farmers and eventually create serious reluctance to crop insurance.

Create a risk pool. This is a key component of a proposed pilot, and its effective management is equally important. A pool needs initial capitalization, and the Government of Nepal should explore the arrangement of financial resources internally and leverage investments from existing insurance companies and the private sector. Indeed, the pool can succeed with strong political commitment. As such, to seed the pool, government should source finance from multilateral agencies such as ADB, the World Bank, KfW, among others, as equity participation or for grants or loans. Part of disaster relief funds allocated by government can also be used for initial funding of the pool.

Align stakeholders' goals. Each stakeholder has different expectations and apprehensions. It is essential that all major stakeholder expectations—government, the insurance sector, and financial institutions— are documented and addressed. This can be achieved by active campaigning, workshops, and other similar tools. Active participation of multilateral organizations, research organizations, industry experts, and nongovernment organizations could help achieve the objective. Agriculture insurance literacy programs and training will increase awareness among farmers.

Equip the stakeholders in Nepal with the technical and operational aspects of disaster insurance and crop insurance. Concepts such as parametric insurance and use of remote sensing for insurance are still new for the country.

Ensure sufficient data infrastructure and quality. These are critical for product design for insurability, and poor data quality is a serious issue. Weather and satellite data are required to price the risks and provide robust triggers and rates for parametric solutions. Government agencies such as the National Disaster Risk Reduction and Management Authority, the Ministry of Geology, the Ministry of Agriculture, the Meteorological Department, and multilateral agencies may be able to provide the necessary data. And automatic weather stations can be established to fulfil additional need.

Provide premium subsidies. These are essential for the success and scalability of the pilot. Potential sharing of premiums among federal, provincial, municipal, individual business, and households, and funding from multilateral organizations, are required to be arranged and worked out. Subsidies may be reduced with time and performance of the pilot.

127.   Explore potential partnerships with organizations such as the International Centre for Integrated Mountain Development and development partners such as the United Nations Capital Development Fund, GIZ, etc. This will increase efficiency and reduce the cost of the pilot.

Use the latest technologies to reduce cost and enhance take-up rates. Digital platforms and smartphone apps may help automate cumbersome insurance processes. Smartphones can also be used, such as for various alerts like weather-related advisories.

Link farm loans with insurance policies (as a policy initiative). Incentives can also be extended to encourage joint bank accounts, with one female family member to pay a share of the premium and receive claims.

Equip female volunteers (*Beema Bahinis*) with smartphones and train them as technical assistants for insurance literacy, documentation, onboarding of farmers, and monitoring. This will help address gender issues.

Monitor regulatory issues in approval of products, setting up a pool, etc. This is needed especially given the volatile political situation. Insurance related regulations are evolving in Nepal.

Have precise understanding of the cost of the pilot to improve planning and implementation.

PARAMETRIC INSURANCE

Parametric insurance offers a way to protect against the financial impact of events that are difficult to predict or quantify using traditional insurance methods. This type of insurance pays out a predetermined amount of money when a predefined event occurs. Unlike traditional insurance policies, which indemnify the policyholder for the actual losses incurred, parametric insurance pays out based on the occurrence of a specified event, such as a hurricane, earthquake, or other disaster, rather than on the actual damage caused by the event.

Parametric insurance policies are often structured as contracts that specify the trigger event, the payout amount, and the parameters that determine whether the trigger event has occurred. These parameters may be based on data from weather stations, seismic sensors, or other sources, and are typically designed to provide a reliable and objective measure of the occurrence of the event. Parametric insurance can provide faster payouts than traditional insurance policies (Table A1.1). This can be particularly valuable in situations where rapid response is needed, such as in the aftermath of a disaster or other catastrophic event.

Table A1.1: Key Differences Between Traditional and Parametric Insurance

	Parametric Policy	Indemnity Policy
Trigger	• Specified parametric threshold exceeded • Occurrence of an event	• Occurrence of a claim
Payment	• Predetermined amount	• Reimbursement of actual losses
Basis risk[a] source	• Modeling inaccuracies • Poor correlation of the index with losses	• Policy conditions, deductibles, and exclusions
Moral hazard	• Low	• Optimized with deductibles and exclusions
Claim adjustment process	• Based on specified parameter • Quick payout	• Based on loss adjustor's assessment • Can be a complex and lengthy process

[a] For parametric insurance, if the payout index and trigger are not aligned with underlying risk exposure, losses of the insured and the insurance payment may not match, known as basis risk. This exists in indemnity insurance too. However, basis risk tends to be greater for parametric solutions as it requires clients deeply understand their exposure to the peril in question.

Source: Swiss Re Institute. Comprehensive Guide to Parametric Insurance (p. 8). https://corporatesolutions. swissre.com/dam/jcr:0cd24f12-ebfb-425a-ab42-0187c241bf4a/2023-01-corso-guide-of-parametric-insurance.pdf.

The many types of parametric insurance policies are designed to cover a wide range of risks:

- Weather insurance: Pays out a predetermined amount of money if specific weather conditions occur, such as excessive rainfall, high winds, or extreme temperatures. Weather insurance can be used by businesses, event organizers, and other entities that are vulnerable to weather-related risks.

- Earthquake insurance: Pays out a predetermined amount of money if an earthquake of a certain magnitude occurs within a specified area. This can be particularly useful in earthquake-prone areas where traditional insurance may be difficult to obtain or prohibitively expensive.

- Crop insurance: Pays out a predetermined amount of money if specific crop yield or quality targets are not met due to weather-related events such as drought or excessive rainfall. This can help farmers manage risk and protect their incomes in the event of crop losses.

- Pandemic insurance: Pays out a predetermined amount of money if a pandemic of a specific severity occurs, such as the coronavirus disease (COVID-19) pandemic. Pandemic insurance is a relatively new type of parametric insurance that has gained popularity in the wake of the global health crisis.

MESO-LEVEL INSURANCE

Meso-level insurance is a type of group- or community-based insurance. It provides coverage to a group of individuals or communities rather than individual policyholders. Policies are designed to address the unique needs of groups that may not have access to traditional insurance products or may face barriers to obtaining coverage.

Meso-level insurance is often used in developing countries, where traditional insurance products may not be available or affordable. These kinds of insurance policies are typically offered by local governments, nongovernment organizations, community-based organizations, and other groups that work closely with the communities they serve. These policies may be tailored to the specific needs and risks of the group, such as coverage for crop losses, livestock mortality, or health-care expenses. For example, microfinance institutions may offer meso-level insurance policies to groups of small farmers or entrepreneurs to help manage risks and protect against unexpected losses.

- The Bangladesh Index-Based Flood Insurance program, launched in 2013 in partnership with Swiss Re, Oxfam, and Institute of Water Modeling Bangladesh. The program provided insurance coverage to vulnerable households and farmers living on river-islands in the Sirajganj District of Bangladesh, using a parametric insurance model that pays out based on the severity of the flood. The program used data generated by flood models to estimate the extent of flooding in the insured areas, and payouts are triggered when floodwaters rise above a certain threshold. The insured was a local nongovernment organization, Manab Mukti Sangtha, and the end beneficiary were the households associated with them.

- Index-Based Livestock Insurance is a meso-level insurance program launched in Kenya in 2010 and developed by the International Livestock Research Institute in collaboration with several partners, including the Kenyan government. The program provides insurance coverage to pastoral communities vulnerable to losses due to drought and other weather-related events. It uses a parametric insurance model in which satellite imagery and weather data are used to estimate the severity of drought in the insured areas. When the estimated drought index exceeds a certain threshold, payouts are triggered, providing financial support to the insured pastoral communities.

NOTE ON PARAMETRIC LANDSLIDE INSURANCE

Nepal is highly vulnerable to landslides due to the inherently unstable nature of mountain areas.
The steep slopes, unstable geography, and intense monsoon rains combine to make the country extremely
hazard prone. Population pressure and improving accessibility by road is leading to human settlement
in hazard-prone areas, placing more people and assets in harm's way (Dahal 2012). A United Nations
University Institute for Environment and Human Security study on the 2014 landslide in Sindhupalchok
District estimated that households with annual income of less than $1,000 incurred median losses of around
$6,000, whereas respondents with an annual income of more than $2,000 had median losses of more than
$10,000 (van der Geest and Schindler 2016). Landslides often also impact community infrastructure,
such as roads, schools, government buildings, and irrigation canals. The worst landslides can cause extreme
consequences, such as creating landslide dams, leading to losses upstream or, in the case of dam bursts,
flash floods downstream.

**Despite high demand, this report believes that traditional property indemnity-based landslide
insurance would be difficult to scale up in Nepal.** This is true for several reasons, including high
administrative expenses (high cost of distribution and loss assessment) and lengthy loss adjustment
procedures.[1] Moreover, Nepal's landslide loss data are limited (WRRDC 2020), which makes it difficult
to model the risk and thereafter secure re-insurance. Future studies should explore (i) precipitation
index-based insurance[2] and (ii) satellite imagery-based insurance for covering landslides.

Several factors directly or indirectly cause slope instability, leading to landslides. The primary causes
(or long-lasting causes) of failure/landslides are inherent in the constituent rock and soil, such as the
force of gravity (slope), rock and soil type and strength, rock structure, soil depth, soil porosity, and soil
permeability. Secondary causes of landslides, either variable or short-lived, are seismic activities, intensity
of precipitation, land use, rock and soil weathering conditions, influence of gullies, and groundwater
conditions. While the primary causes determine the inherent vulnerability, the secondary causes such
as rainfall and earthquakes are the main factors that determine the frequency of landslides within a
limited observation period. The rainfall in Nepal is essentially controlled by monsoon winds and the
physical geography. Approximately 80% of total annual rainfall occurs between June and September.
Average annual rainfall is 1,600 millimeters (mm), but varies by eco-climatic zones, such as 3,345 mm in
Pokhara and below 300 mm in Mustang. It is not uncommon in Nepal for 10% of total annual precipitation
to occur in a single day or 50% of the total during 10 days of the rainy season. Such heavy rainfall events play
an important role in triggering landslides.

---

[1]   Loss surveys in settlements (the primary target for landslide insurance) is going to be expensive and time-consuming.

[2]   The advantages of parametric or index-based insurance are mentioned in Appendix 1. Muñoz-Torrero Manchado et al.
      (2021, p. 6).

Scope of precipitation index-based landslide insurance: Numerous studies have explored the strong correlation between the amount and intensity of rainfall and frequency of landslides in Nepal, which allows for a possibility of developing precipitation index-based insurance for landslide. The primary causes (or long-lasting causes) of failure/landslides are inherent in the constituent rock and soil, such as the force of gravity (slope), rock and soil type and strength, rock structure, soil depth, soil porosity, and soil permeability. Secondary causes of landslides, either variable or short-lived, are seismic activities, intensity of precipitation, land use, rock and soil weathering conditions, influence of gullies, and groundwater conditions. While the primary causes determine the inherent vulnerability, the secondary causes such as rainfall and earthquakes are the main factors that determine the frequency of landslides within a limited observation period. Approximately 80% of total annual rainfall occurs between June and September. The average annual rainfall is 1,600 mm, but it varies by eco-climatic zones, such as 3,345 mm in Pokhara and below 300 mm in Mustang. It is not uncommon in Nepal for 10% of the total annual precipitation to occur in a single day or 50% of the total during 10 days of the rainy season (Dahal 2012). Such heavy rainfall events play an important role in triggering landslides. Landslides show high correlation with both total monsoon precipitation as well as with shorter periods of 10 days of accumulated precipitation. The highest incidence of landslides occurs during warm monsoons followed by an especially humid monsoon, with correlation at r = 0.77 (Muñoz-Torrero Manchado et al. 2021). Support for identifying landslide-triggering points is essential; however, other geotechnical factors must also be considered during the design process. A detailed study will be necessary to address these aspects comprehensively.

Data: Most precipitation data in Nepal are from stations/observatories in the valley regions and may not represent the actual precipitation in the adjust slopes/ridges. Also, historical loss data due to landslides is limited, as pointed out by The Water Resources Research and Development Centre, under the Ministry of Energy, Water Resources, and Irrigation, which makes it difficult to model the losses. Future studies should consider (i) setting up automated weather stations/automated rain gauges in pilot locations and upper slopes for more accurate precipitation data, and (ii) investing in creating loss estimation models.

Considerations for pilot: Future studies/pilots should consider only those locations for pilots which have the correlation between precipitation and landslides. Future studies should note that the Water Resources Research and Development Centre, under the Ministry of Energy, Water Resources, and Irrigation, has developed a rainfall hazard assessment model[3] that uses rainfall and landslide susceptibility map. This tool might provide the necessary platform to build the parametric insurance scheme.

Index-based insurance carries basis risk as the precipitation and landslide intensity are not perfectly correlated. The pilot should invest in adequate monitoring and dispute resolution mechanisms including use of remote sensing-based monitoring and community-based reporting. The client should note that a number of organizations are developing remote-sensing-based landslide monitoring and forecasting system for Nepal, such as the International Centre for Integrated Mountain Development's SERVIR-HKH Initiative, a joint initiative of United States Agency for International Development and the National Aeronautics and Space Administration (NASA), which is working on developing a landslide monitoring and forecast system. Monitoring systems based on Synthetic Aperture Radar should also be explored.

---

[3]   Earth Engine App for Landslide Hazard Assessment Model. This is developed by Water Resources Research and Development Centre, the Ministry of Energy, Water Resources and Irrigation, and the Government of Nepal. The tool uses the Landslide Hazard Assessment model for Situational Awareness created at the Goddard Space Flight Center, NASA to identify potential landslide areas. https://wrradc.users.earthengine.app/view/lhamnepal.

EXPANDING AUTOMATED WEATHER STATIONS FOR INSURANCE

In Nepal, the Department of Hydrology and Meteorology (under the Ministry of Energy, Water Resources, and Irrigation), the primary meteorological agency, manages several surface observation stations across the country.[1]

This study recommends that future pilots/studies should attempt to develop the weather index-based crop insurance products around the department's surface observation stations because it is easier to gain farmer's trust with government data. But if the target locations do not have the department's surface observation stations, automated weather stations (AWS) might need to be erected to record weather data. The AWS is a "meteorological station which takes the meteorological data at regular pre-set intervals and transmits it automatically." These machines have built-in electronic sensors to record the data, which is then transmitted over GPRS or satellite to a data-agency managed central server (Table A4.1).

Table A4.1: Automatic Weather Station Advantages and Disadvantages

Advantages	Disadvantages
Data are recorded automatically, transmitted at the same time throughout the network on a real-time basis, and available outside normal working hours.	Possibility of loss of the data for some of the meteorological parameters, e.g., amount of cloud, sudden weather fluctuations
Automatic weather stations can be established in remote locations and the health of stations can be monitored at a centralized location.	Loss of the data due to sabotage (theft of solar panel, battery, etc.)
The homogeneity of networks can be ensured by standardizing the measuring techniques	If an automatic weather station stops working, data are not available for a period.
Lowers the operational costs by reducing the number of observers	Maintenance can become a challenge, especially when rain buckets get clogged.

Source: Author.

Weather index-based crop insurance works on the assumption that there is a functional correlation between yield and underlying weather parameters, and uses deviation of weather parameters as the proxy for the crop yield losses for calculating insurance compensation. While a weather observation station or instrument captures weather information for its location only (point information only), it is assumed that this data represent the weather conditions of a radius of 10–20 kilometers depending on the topography of the plains, and depending upon technology integrated AWS collects varied weather parameters (Table A4.2).

---

[1]   A map of surface observation stations maintained by the Department of Hydrology and Meteorology can be accessed at https://www.dhm.gov.np/hydrology/surface-observation.

Table A4.2: Parameters Observed by Automatic Weather Stations

Main Weather Parameters Observed by an AWS	Additional Parameters That Can Be Observed by an AWS
• Air temperature	• Visibility
• Relative humidity	• Soil moisture and
• Atmospheric pressure	• Soil temperature
• Rainfall	• Solar radiation
• Wind speed	• Leaf wetness
• Wind direction	

AWS = automatic weather station.

Source: Niubol. Introduction and Function of Automatic Weather Index. https://www.niubol.com/Product-knowledge/Introduction-to-automatic-weather-stations.html#:~:text=It%20usually%20consists%20of%20sensors,speed%2C%20wind%20direction%20and%20precipitation.

As the Department of Hydrology and Meteorology is primarily a research body concerned with national climatological research, it might be challenging for the department to manage new surface observation stations. **This report strongly recommends this gap be filled by encouraging private weather data providers, which would serve insurance companies for a fee.**

Before engaging third-party data providers, certain standards in the AWS set up need to be stipulated, and maintenance and data accuracy ensured. The government should set up a committee under the leadership of Department of Hydrology and Meteorology that should draft the guidelines covering (i) AWS equipment standards, (ii) AWS installation and maintenance standards, (iii) data quality standards, and (iv) guidelines for third-party accreditation and data certification services. The following sections aim to initiate discussion of guidelines.[2]

Weather stations for weather index-based crop insurance need not maintain the same level of specifications and standards as the World Meteorological Organization for climatological and forecasting. It is often challenging to find secure locations that meet the organizations climatological data standards for erecting automatic weather station standards. As such, relaxation of guidelines is needed for weather data collection for weather insurance.

In Nepal, because most automatic weather stations will be installed remotely, sensors must be robust, fairly maintenance free, and have no intrinsic bias or uncertainty in how they sample variables to be measured. The quality check of AWS data needs to be monitored at the data processing center server, as per the norm of World Meteorological Organization norms (i.e., in parity, range, and consistency checks; comparison of the data with nearby stations; comparison of the data with climatological average). Quality control assures quality and consistency of data output through carefully designed procedures focused on good maintenance practices, repair, calibration, and data quality checks.[3]

---

[2]   This section borrows heavily from the AWS-related guidelines proposed by Government of India.

[3]   Guidance on automatic weather systems and their implementation can be seen on the World Meteorological Organization website at https://library.wmo.int/index.php?lvl=notice_display&id=11269#.Y48c2ctBxnJ. The manual on instrumentation and operations for AWS for agrometeorological application is at https://library.wmo.int/index.php?lvl=notice_display&id=1694#.Y48dXctBxnl.

It is important to consider the lifetime costs of an AWS rather than simply the initial cost. Generally, the lower the initial cost, the higher the ongoing cost to maintain acceptable data. In the end, this may result in either a higher total cost or long periods with no useful data. Lifetime cost includes (i) cost of the AWS, (ii) cost of installation and peripherals, (iii) operation costs, (iv) preventive and corrective maintenance, and (v) software and server costs.

> Rain data are typically recoded using a tipping bucket rain gauge, which basically collects the rain in bucket-like structure and passes the collected water through a sensor that measures the volume of water. If the rain bucket gets clogged with leaves or dirt, then the device will fail to record or record erroneous data. It is important to clean the rain bucket at regular intervals.

To ensure minimum criteria/benchmark for weather data, the Department of Hydrology and Meteorology should also consider certification of weather data and AWS by third-party agencies. The certification vendor can be chosen through competitive bidding. A few requirement specifications for possible third-party certifying agencies organizations include:

- adequate resources and personnel to audit AWS;
- proven experience in weather data quality checks;
- facility to check data quality per the World Meteorological Organization or Department of Hydrology and Meteorology guidelines;
- expertise in developing algorithms for weather data quality checks and/or certification;
- exposure and experience of other data networks available in the market, including meteorological satellites; and
- exposure and experience in meteorological satellites and experience in handling weather data from those satellites.

INSURANCE POOL EXAMPLES

A. Spanish Agriculture Insurance System

Spain has more than 47% of its territory, and more than 2.2 million people are employed, in agricultural holdings. To protect against agriculture risk and for farmers' financial stability, the government developed the Spanish Agriculture Insurance System, a robust risk transfer tool. The system offers a legal agreement among all political groups, which makes it stable and features long-term government commitment to farmers and insurance companies. Primary features include the following:

- It is a public–private system that allows farmers to transfer their farm risks to the insurance companies. Insurance company participation is through Agroseguro.

- Agroseguro is an association of insurance companies willing to participate in the annual agricultural insurance program. It is a joint stock company that offers share capital in equal proportion to risk coverage.

- Agroseguro functions as co-insurance pool.

- It covers crop, livestock, aquaculture, and forestry risks.

- Risks covered are the damage caused by hail, fire, drought, frost, flood, hurricane or warm wind, snowfall, excess humidity, other climatic adversities, accidents, pests and diseases to agricultural production and, optionally, installations. This risk protection is associated with multi-peril policies.

- Participation is voluntary for farmers.

- Those who subscribe to an agricultural insurance policy are obliged to insure all production of the same class in the national territory.

- Insurance policies are subsidized and supervised by the government. Regional government also provides subsidies.

- The insured under the Spanish Agriculture Insurance System cannot be compensated otherwise. No disaster relief payments for insurable losses.

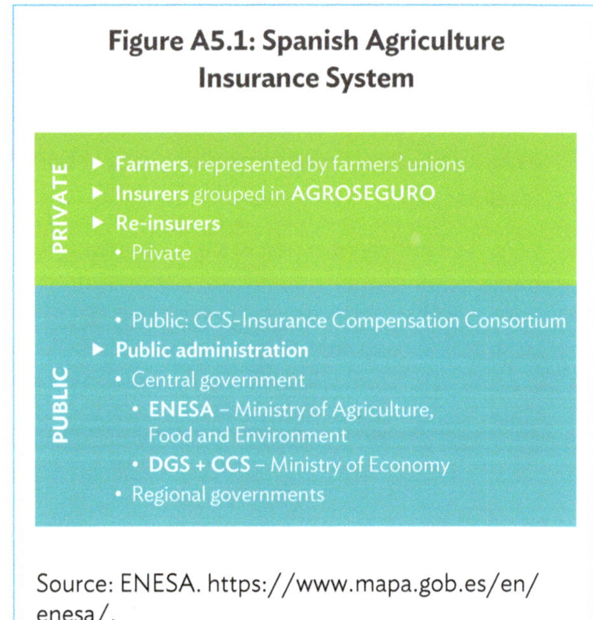

Figure A5.1: Spanish Agriculture Insurance System

Source: ENESA. https://www.mapa.gob.es/en/enesa/.

B. Ghana Agriculture Insurance Pool

The Ghana Agricultural Insurance Programme is a government-led initiative designed to help farmers manage risks in agricultural production, including crop failure, disasters, and other unforeseen events that can affect yields and income. The program is implemented by the National Insurance Commission and the Ministry of Food and Agriculture, with support from the World Bank and other development partners. The program uses a combination of insurance and risk management strategies to protect farmers from the financial losses associated with crop failure or other risks.

Farmers who participate pay a premium based on the value of their crops, and in return, they receive insurance coverage for losses due to disasters, pests and diseases, and other risks. The insurance coverage is designed to provide farmers with financial compensation that can help them recover from loss and continue farming.

The Ghana Agricultural Insurance Programme is part of a broader effort to promote agricultural development and food security in the country. By providing farmers with access to insurance services, the program aims to increase productivity, reduce poverty, and promote sustainable agriculture.

C. Turkish Catastrophe Insurance Pool

Türkiye is among the world's most seismically active regions. The Turkish Catastrophe Insurance Pool was established in 2000 with support of the Government of Türkiye and the World Bank, after the 7.4-magnitude Marmara earthquake in 1999 that killed 18,373 people, according to official figures. The Turkish Industrialist and Businessmen Association claims the economic cost was around $17 billion (BBC News Turkey 2019).

The Turkish Catastrophe Insurance Pool provides coverage against disasters caused by natural hazards such as earthquakes, floods, landslides, and fires. It is a nonprofit organization owned by the Turkish government and regulated by the Turkish Catastrophe Insurance Regulation and Supervision Agency. It provides insurance coverage to residential and commercial properties, including apartments, houses, and businesses, against disasters. The damage covered includes loss of income, temporary accommodation expenses, and emergency repair costs.

The Turkish Catastrophe Insurance Pool is mandatory, which means all property owners are required to have insurance coverage against disasters. Property owners can obtain coverage from private insurance companies or through the pool. It is funded by premiums paid by property owners and the Turkish government. The premiums are calculated based on location, construction type, and size of the property. The pool also receives financial support from the Turkish government as subsidies and reinsurance.

The insured amounts are limited to 640,000 Turkish lira ($33,000) per dwelling. Sums above this amount may be underwritten by insurers on a facultative basis. A 2% deductible of the insured sums is applied for every claim.

The insurance companies authorized to conclude the Compulsory Earthquake Insurance contracts on behalf of the Turkish Catastrophe Insurance Pool are paid a commission of 12.5% for the risks located in Istanbul and 17.5% for risks located in other provinces (DASK 2022).

Earthquakes and other disasters are frequent in Türkiye, and the Catastrophe Insurance Pool has successfully paid out over $2 billion (SteelRadar 2023) in claims since its establishment. The pool has paid out $340.36 million in claims for the 7.8-magnitude Türkiye–Syria earthquake of 6 February 2023 (Willard 2023).

Within the scope of the technical operator agreements entered into with the Ministry of Treasury and Finance, which is responsible for the surveillance and supervision of the pool, Milli Reasürans served as the technical operator from 2000 to 2005, and Eureko Sigorta from 2005 to 2020. As of 2020, the pool has been managed by Türk Reasürans A.Ş., which is affiliated to the Ministry of Treasury and Finance, as the technical operator (DASK 2022).

NOTABLE CROP INSURANCE PILOTS IN NEPAL

Table A6.1: Area Yield Index Insurance (United Nations Capital Development Fund)[a]

Pilot Location	• Province 1 (Morang District)
Stakeholders	• Funding agencies: United Nations Capital Development Fund, Government of Nepal (with premium subsidies). • Implementation agencies: Jeevan Bikas Laghubitta Bittiya Sanstha Limited (Jeeva Bikas, a local microfinance company), Pula Advisors (an agriculture insurance consultancy and broker).
Project Numbers	• Farmers insured: 15,380 • Total Insured: $4,000,000
Product Details	• Coverage period: monsoon paddy. • Index: Estimated yield for the insured unit area.

Index-Based Flood Insurance (United States Agency for International Development)

Pilot Location	• Ward 3, Bhajani Municipality
Stakeholders	• Funding: USAID • Implementation: Sagarmatha Insurance Company • Target beneficiary: Consortium for Land Research and Policy Dialogue took out a group insurance policy on behalf of 100 farmers.
Product Details	• Coverage Period: 1 August 2022 to 25 November 2022 • Risk: Flood • Crops insured: Paddy • Sum Insured: NRs3,568,045.4 • Premium: NRs183,572.28 (5% of sum insured)

NRs = Nepalese rupees, USAID = United States Agency for International Development.

[a] United Nations Nepal. 2023. Bringing Agricultural Insurance to Climate-Vulnerable Farmers: A Unique Pilot Program in Nepal Shows How to Unlock the Benefits of Index-Based Insurance for Smallholders. https://nepal.un.org/en/225296-bringing-agricultural-insurance-climate-vulnerable-farmers-unique-pilot-program-nepal-shows.

Source: Project brochure.

Table A6.2: Policy Triggers and Payoff Formula for Index-Based Flood Insurance

Flood Water Depth (in meters)	Damage Calculated Based Flood Days				Payout (Cost of Input During Vegetative Stage (Ha)	Payout on Yield Decline (Ha)
	1–2 Days	3–5 Days	6–7 Days	>7 Days		
Below 0.6	ND	ND	ND	ND	ND = 0	ND = 0
0.6 to 0.8	ND	LD	LD	MD	LD = 5,000 MD = 10,000	LD = 18,750 MD = 37,500
0.8 to 1	ND	LD	LD	MD	LD = 5,000 MD = 10,000 HD = 15,000	LD = 18,750 MD = 37,500 HD = 56,250
>1	ND	LD	MD	HD	LD = 5,000 MD = 10,000 HD = 15,000	LD = 18,750 MD = 37,500 HD = 56,250

ha = hectares, HD = High Damage, LD = Low Damage, MD = Medium Damage, No = No Damage.

Source: Project brochure.

Table A6.3: Index-Based Flood Insurance (Stonestep)

Pilot Location	• Sudurpaschim province (Tikapur, Rajapur, and Madhuban municipalities) • Lumbini provinces (Janaki and Geruwa Rural municipalities)
Stakeholders	• Funding agencies: InsuResilience Solutions Fund and Zurich Flood Resilience Alliance • Implementation: Practical Action, Global Parametrics, Sikhar Insurance, and Stonestep • Targeted beneficiary: Smallholder paddy farmers residing in flood-prone areas of downstream communities of the Karnali River. Farmers are insured under group insurance policies via farmers' cooperatives.
Product Details	• Cover Period: Monsoon period (June to October) • Risk: Fluvial flood (Karnali River) • Crops insured: Paddy • Sum Insured: NRs3,600 per Kattha (where 1 hectare = 29.586 Kattha) • Premium: 7% of sum insured

NRs = Nepalese rupees.

Source: Stonestep brochure.

Table A6.4: Policy Triggers and Payoff Formula for Stonestep's Index-Based Flood Insurance

Water Level (meters) (in index number 280, Chisanpani Karnali)	Payoff (% of sum insured)
10.80 to 11.80	10
11.81 to 12.80	25
12.81 to 13.80	50
13.81 to 14.80	75
14.81 and above	100

Source: Stonestep project brochure.

GOVERNMENT-BACKED EARTHQUAKE INSURANCE SCHEMES

The New Zealand Earthquake Commission is a government entity providing insurance to residential property owners for damage to houses and contents stemming from earthquakes, landslides, volcanic eruption, hydrothermal activity, or tsunamis. It also provides storm and flood coverage for residential land that allows property access or that includes building platforms. The commission transfers the financial risk posed by New Zealand's natural hazards through financial arrangements, including (i) the Natural Disaster Fund, (ii) an international reinsurance program renewed every year, and (iii) a backstop government guarantee in the event that the New Zealand Earthquake Commission's reserves and reinsurance lines are exhausted (under Section 16 of the Earthquake Commission Act 1993). The Treasury may meet the deficiency of funds by providing either a grant or a loan.

France's CATNAT insurance scheme, a public-private partnership, has been the backbone of disaster recovery financing in France since its establishment in 1982. The scheme has been put in place to provide insurance for hazards otherwise considered "uninsurable", i.e., such as flooding, avalanches, volcanic activity, or earthquakes. Both private and public assets can be covered by hazard insurance via the scheme. Funding for CATNAT comes from an additional premium, fixed by the state at a mandatory uniform rate, for all property and motor vehicle insurance policies. To prevent illiquidity in case a major disaster triggers insurance payouts beyond available reserves, CATNAT is backed by a state guarantee. If claims exceed 90% of the special reserve and annually defined equalization reserves, the government is required to step in.

Japan Earthquake Reinsurance (JER) offers insurance through the private insurance market, the government of Japan has a key role in retaining a portion of the liability.[1] Under the scheme, the private and public sectors share the aggregate limit of indemnity for a single seismic event (up to ¥12 trillion) as follows:

- For liabilities up to ¥125.9 billion, the JER is liable for 100% of insurance claims.

- For amounts over ¥125.9 billion and up to ¥266.1 billion, the central government is liable for 50% and the JER and private insurers (i.e., those to which the JER has retroceded risk) are liable for 50%.

- For amounts from ¥224 billion to ¥12 trillion, the central government is liable for approximately 99.8%. JER and private insurers are liable for approximately 0.2%.

- If earthquake insurance liabilities for one event exceed the indemnity cap of ¥11.3 trillion, the government can decide to provide additional resources on a need basis.

---

[1] Details of JER can be accessed at https://www.nihonjishin.co.jp/pdf/disclosure/english/2021/en_05.pdf.

REFERENCES

Amarnath, G., R. P. S. Malik, and A. Taron. 2021. Scaling up Index-based Flood Insurance (IBFI) for agricultural resilience and flood-proofing livelihoods in developing countries. Vol. 180. International Water Management Institute (IWMI).

Asian Development Bank (ADB). 2019a. *Nepal: Flood Risk Sector Assessment.* Consultant's report (TA9634-REG). https://www.adb.org/sites/default/files/project-documents/52014/52014-001-dpta-en.pdf.

―――. 2019b. The Enabling Environment for Disaster Risk Financing in Nepal—A Country Diagnostics Assessment.

―――. 2024. Nepal Macroeconomic Update. https://www.adb.org/sites/default/files/institutional-document/1001006/nepal-macroeconomic-update-202409.pdf.

BBC News Turkey. 2019. August 17 Earthquake: What Happened in 1999 and Later, How Many People Lost Their Lives? (in Turkish). 12 August. https://www.bbc.com: https://www.bbc.com/turkce/haberler-turkiye-49322860.amp.

Bhandari, D., S. Nepuane, P. Hayes, B. Regmi, and P. Marker. 2020. Oxford Policy Management "Disaster Risk and Management in Nepal: Delineation of Role and Responsibility." https://www.opml.co.uk/files/Publications/a1594-strengthening-the-disaster-risk-response-in-nepal/delineation-of-responsibility-for-disaster-management-full-report-english.pdf.

Budhathoki, N. K. 2019. Farmers' Interest and Willingness-to-Pay for Index-Based Crop Insurance in the Lowlands of Nepal. *Land Use Policy.* 85. pp. 1–10. https://researchers.cdu.edu.au/en/publications/farmers-interest-and-willingness-to-pay-for-index-based-crop-insu.

Central Bureau of Statistics (CBS). 2021. National Population and Housing Census 2021. National Statistics Office, Government of Nepal. Kathmandu. https://censusnepal.cbs.gov.np/results/files/result-folder/National%20Report_English.pdf.

Dahal, R. K. 2012. Rainfall-Induced Landslides in Nepal. *International Journal of Japan Erosion Control Engineering.* 5(1). https://www.jstage.jst.go.jp/article/ijece/5/1/5_1/_pdf.

DASK. 2022. Tariff. Turkish Natural Catastrophe Insurance Pool. https://www.dask.gov.tr/en/tariff.

Dhakal, M. 2023. NIA Recommends Payment via Local Govt. *The Rising Nepal.* 3 June. https://risingnepaldaily.com/news/27555.

Gallin, L. 2023. Swiss Re CorSo and Aon Deliver Parametric Cover to Nepal Renewable Energy Project. *Arthemis.* 3 March. https://www.artemis.bm/news/swiss-re-corso-and-aon-deliver-parametric-cover-to-nepal-renewable-energy-project/.

Ghimire, L. B. 2017. 500 Industries Shut Down in Morang After Flooding. *The Kathmandu Post.* 16 August. https://kathmandupost.com/national/2017/08/16/500-industries-in-morang-shut-down-following-floods.

Government of Nepal. 2015. The Constitution of Nepal. Kathmandu. https://ag.gov.np/files/
    Constitution-of-Nepal_2072_Eng_www.moljpa.gov_.npDate-72_11_16.pdf.

Guo, W. 2016. Farmers' Perception of Climate Change and Willingness to Pay for Weather-Index Insurance
    in Bahunepati, Nepal. Working paper. https://digitalrepository.unm.edu/cgi/viewcontent.
    cgi?article=1073&context=hprc.

Himalayan Everest Insurance Limited. n.d. Crop Insurance. https://hei.com.np/product/crop-insurance/.

Investopaper. 2023. Current Status of Insurance Business in Nepal. 25 September.
    https://www.investopaper.com/news/current-status-of-insurance-business-in-nepal/.

Karki, R., R. Talchabhadel, and S. K. Baidya. 2016. New Climatic Classification of Nepal. *Theoretical and
    Applied Climatology*. 125: pp. 799–808.

MacAskill, A. and G. Sharma. 2015. Four Months after Quakes, Nepal Fails to Spend Any of $4.1 Billion
    Donor Money. *Reuters*. 2 September. https://www.reuters.com/article/us-nepal-rebuilding-
    idUSKCN0R20GU20150902.

Muñoz-Torrero Manchado, A., S. Allen, J. A. Ballesteros-Cánovas, A. Dhakal, M. R. Dhital, and M. Stoffel.
    2021. Three Decades of Landslide Activity in Western Nepal: New Insights into Trends and Climate
    Drivers. *Landslides*. 18(6): pp. 2001–2015.

Nepal Agriculture Economics Society. 2021. Effectiveness and Suitable Modality of Crop Insurance for
    Bagmati Province, Nepal. Ministry of Land Management Agriculture and Cooperatives Bagamati
    Province, Hetauda, Agriculture Development Directorate. https://doad.bagamati.gov.np/sites/
    default/files/final%20report_Crop%20insurance_NAES_2078_TBhandari_0.pdf.

Nepal Rastra Bank. 2024. Nepal Rastra Bank "Economic and Financial Status of Country."
    https://www.nrb.org.np/contents/uploads/2024/08/Current-Macroeconomic-and-Financial-
    Situation-Nepali-Based-on-Annual-data-of-2080.81-2.pdf.

New Business Age. 2021. Making Agriculture Insurance Meaningful. 21 December.
    https://www.newbusinessage.com/MagazineArticles/view/3170.

Pandey, A. 2022. Credit and Financial Access in Nepalese Agriculture: Prospects and Challenges.
    The Journal of Agriculture and Environment. 23. pp. 56–70.

Praseed Thapa, A. A. 2018. Importance, Scope, and Status of Agriculture Insurance in Nepal.
    Journal of Agricultural Economics and Rural Development. 4(1). pp. 365–371.

Reinsurance News. 2021. New Partnership Establishes Parametric Solution for Nepalese Farmers. 1 April.
    https://www.reinsurancene.ws/new-partnership-establishes-parametric-solution-for-nepalese-
    farmers/.

Schneeberger, K., M. Huttenlau, B. Winter, T. Steinberger, S. Achleitner, and J. Stötter. 2019.
    A Probabilistic Framework for Risk Analysis of Widespread Flood Events: A Proof-of-Concept Study.
    Risk Analysis. 39(1): pp. 125–139.

SteelRadar. 2023. Claim Payment of TCIP Amounted to More than ₺2 Billion. https://www.steelradar.
    com/en/haber/tcips-claim-payment-amounted-to-more-than-2-billion/.

Thapa, B. K. 2021. Agricultural Credit: Problems and Prospects. *The Rising Nepal.* 5 August. https://old.risingnepaldaily.com/opinion/agricultural-credit-problems-prospects.

The Kathmandu Post. 2015. Major Revenue Sources Under Central Govt. 22 September. https://kathmandupost.com/money/2015/09/22/major-revenue-sources-under-central-govt.

———. 2017. Flood Damage Insurance Claims Total Rs2.5 billion. 15 September. https://kathmandupost.com/money/2017/09/15/flood-damage-insurance-claims-total-rs25-billion.

———. 2018. Ilam Farmers' Response to Agro Insurance 'Lukewarm'. 26 February. https://kathmandupost.com/money/2018/02/26/ilam-farmers-response-to-agro-insurance-lukewarm.

Trendafiloski, G. 2020. The Seismic Impact on the Protection Gap: Five Years After the Gorkha Earthquake in Nepal. AON UK. https://www.aon.com/reinsurance/thought-leadership/20200422-if-nepal-prot-gap.

United Nations Development Programme (UNDP). 2022. Nepal Moves Up One Place in Human Development, Ranks 143rd. https://www.undp.org/nepal/press-releases/nepal-moves-one-place-human-development-ranks-143rd.

van der Geest, K. and M. Schindler. 2016. Case Study Report: Loss and Damage from a Catastrophic Landslide in Sindhupalchok District, Nepal. United Nations University Institute for Environment and Human Security. https://collections.unu.edu/eserv/UNU:5854/CSR_Nepal_Final.pdf.

Water Resources Research and Development Centre (WRRDC). 2020. Preparation of Landslide Catalogue (1970–2019) of Nepal. https://wrrdc.gov.np/storage/listies/December2020/WRRDC-Research_Letter_Issue_2_November.pdf.

Willard, J. 2023. Turkish Catastrophe Insurance Pool Pays $340–4 to Earthquake Victims. *Reinsurance News.* 29 March. https://www.reinsurancene.ws/turkish-catastrophe-insurance-pool-pays-340-4mn-to-earthquake-victims/.

Winrock International. 2021. Expanding Access to Finance for Nepali Farmers. 5 May. https://agrilinks.org/post/expanding-access-finance-nepali-farmers.

World Bank and the Asian Development Bank. 2021. *Climate Risk Country Profile: Nepal.* https://www.adb.org/sites/default/files/publication/677231/climate-risk-country-profile-nepal.pdf.

World Economic Forum. 2015. *Building Resilience in Nepal through Public–Private Partnerships.*